A Readable Introduction to Real Mathematics

Undergraduate Texts in Mathematics

Undergraduate Texts in Mathematics are generally aimed at third- and fourth-year undergraduate mathematics students at North American universities. These texts strive to provide students and teachers with new perspectives and novel approaches. The books include motivation that guides the reader to an appreciation of interrelations among different aspects of the subject. They feature examples that illustrate key concepts as well as exercises that strengthen understanding.

For further volumes:
http://www.springer.com/series/666

Daniel Rosenthal • David Rosenthal
Peter Rosenthal

A Readable Introduction
to Real Mathematics

 Springer

APR 0 7 2017

Daniel Rosenthal
Department of Mathematics
University of Toronto
Toronto, ON, Canada

David Rosenthal
Department of Mathematics
 and Computer Science
St. John's University
Queens, NY, USA

Peter Rosenthal
Department of Mathematics
University of Toronto
Toronto, ON, Canada

ISSN 0172-6056   ISSN 2197-5604 (electronic)
ISBN 978-3-319-37950-0   ISBN 978-3-319-05654-8 (eBook)
DOI 10.1007/978-3-319-05654-8
Springer Cham Heidelberg New York Dordrecht London

Mathematics Subject Codes (2010): 03E10, 11A05, 11A07, 11A41, 11A51, 11-01, 51-01, 97-01, 97F30, 97F40, 97F50, 97F60, 97G99, 97H99

Printed on acid-free paper

Springer is part of Springer Science+Business Media (www.springer.com)

To the memory of Harold and Esther Rosenthal who gave us (and others) the gift of mathematics.

Preface

The fundamental purpose of this book is to teach you to understand mathematical thinking. We have tried to do that in a way that is clear, engaging and emphasizes the beauty of mathematics. You may be reading this book on your own or as a text for a course you are enrolled in. Regardless of your reason for reading this book, we hope that you will find it understandable and interesting.

Mathematics is a huge and growing body of knowledge; no one can learn more than a fraction of it. But the essence of mathematics is thinking mathematically. It is our experience that mathematical thinking can be learned by almost anyone who is willing to make a serious attempt. We invite you to make such an attempt by reading this book. It is important not to let yourself be discouraged if you can't easily understand something. Everyone learning mathematics finds some concepts baffling at first, but usually, with enough effort, the ideas become clear.

One way in which mathematics gets very complex is by building on itself; some mathematical concepts are built on a foundation of many other concepts and thus require a great deal of background to understand. That is not the case for the topics discussed in this book. Reading this book does not require any background other than basic high school algebra and, for parts of Chapters 9 and 12, some high school trigonometry.

A few questions, among the many, that you will easily be able to answer after reading this book are the following: Is $13^{217} \cdot 37^{92} \cdot 41^{15} = 19^{111} \cdot 29^{145} \cdot 43^{12} \cdot 47^5$ (see Chapter 4)? Is there a largest prime number (i.e., a largest whole number whose only factors are 1 and itself) (Theorem 1.1.2)? If a store sells one kind of product for 9 dollars each and another kind for 16 dollars each and receives 143 dollars for the total sale of both, how many products did the store sell at each price (Example 7.2.7)? How do computers send secret messages to each other (Chapter 6)? Are there more fractions than there are whole numbers? Are there more real numbers than there are fractions? Is there a smallest infinity? Is there a largest infinity (Chapter 10)? What are complex numbers and what are they good for (Chapter 9)?

The hardest theorem we will prove concerns construction of angles using a compass and a straightedge. (A straightedge is a ruler-like device but without measurements marked on it.) If you are given any angle, it is easy to bisect it (i.e., divide it into two equal subangles) by using a compass and a straightedge (we will show you how to do that). This and many similar results were discovered by the Ancient Greeks. The Ancient Greeks wondered whether angles could be "trisected" in the sense of being divided into three equal subangles using only a straightedge and a compass. A great deal of mathematics beyond that conceived of by the Ancient Greeks was required to solve this problem; it was not solved until the 19th century. It can be proven that many angles, including angles of 60 degrees, cannot be so trisected. We present a complete proof of this as an illustration of complicated but beautiful mathematical reasoning.

The most important question you'll be able to answer after reading this book, although you would have difficulty formulating the answer in words, is: what is mathematical thinking really like? If you read and understand most of this book and do a fair number of the problems that are provided, you will certainly have a real feeling for mathematical thinking.

We hope that you read this book carefully. Reading mathematics is not like reading a novel, a newspaper, or anything else. As you go along, you have to really reflect on the mathematical reasoning that we are presenting. After reading a description of an idea, think about it. When reading mathematics you should always have a pencil and paper at hand and rework what you read.

Mathematics consists of theorems, which are statements proven to be true. We will prove a number of theorems. When you begin reading about a theorem, think about why it may be true before you read our proof. In fact, at some points you may be able to prove the theorem we state without looking at our proof at all. In any event, you should make at least a small attempt before reading the proof in the book. It is often useful to continue such attempts while in the middle of reading the proof that we present; once we have gotten you a certain way towards the result, see if you can continue on your own.

If you adopt such an approach and are patient, we believe that you will learn to think mathematically. We are also convinced that you will feel that much of the mathematics that you learn is beautiful, in the sense that you will find that the logical argument that establishes the theorem is what mathematicians call "elegant."

We chose the material for this book based on the following criteria: the mathematics is beautiful, it is "real" in the sense that it is useful in many mathematical contexts and it is accessible without a great deal of mathematical background. The theorems that we prove have applications to mathematics and to problems in other subjects. Some of these applications will be presented in what follows.

Each chapter ends with a section entitled "Problems." The problems sections are divided into three subsections. The first, "Basic Exercises," consists of problems whose solution you should do to assure yourself that you have an understanding of the fundamentals of the material. The subsections entitled "Interesting Problems" contain problems whose solutions depend upon the material of the chapter and seem to have mathematical or other interest. The subsections labeled "Challenging Problems" contain problems that we expect you will, indeed, find to be quite challenging. You should not be discouraged if you cannot solve some of the problems. However, if you do solve problems that you find difficult at first, especially those that we have labeled "challenging," we hope and expect that you will feel some of the pleasure and satisfaction that mathematicians feel upon discovering new mathematics.

Each chapter is divided into sections. Important items, such as definitions and theorems, are numbered in a way that locates them within a chapter and a section of that chapter. We put the chapter number, then the section number, and then the number of the item within that section. For example, 7.2.4 refers to the fourth item in section two of chapter seven.

Since the only prerequisite for understanding this book is high school algebra, it is suitable as a textbook for a wide variety of courses. In particular, it is our view that it would be appropriate for courses for general arts and sciences students who want to get an appreciation of mathematics, for courses for prospective teachers, and for an introductory course for mathematics majors. Instructors can vary the level of the course by the pace at which they proceed, the difficulty of the problems that they assign, and the material they omit. The book is also written so as to be useable for independent study by anyone who is interested in learning mathematics. In particular, high school students who like mathematics might be directed to this book.

Instructors and readers who wish to omit some of the material (perhaps only at first) should be aware of the following. Chapters 1 through 7 each depend, at least to some extent, on their predecessors. Chapter 8 uses some of the material in Chapter 4. Chapters 9–11 are essentially independent of each other and of all other chapters. Chapter 12 depends basically only on Chapter 11 and on the concepts of rational and irrational numbers as discussed in Chapter 8.

This book was developed from lecture notes for a course that was given at the University of Toronto over a period of 15 years. It has been greatly improved by suggestions from students and colleagues. We are particularly grateful to Professor Heydar Radjavi of the University of Waterloo for his assistance and to two anonymous reviewers for their comments. In spite of all the suggestions, we are sure that further improvements could be made. We would appreciate your sending any comments, corrections, or suggestions to any of the authors at their e-mail addresses given below.

Daniel Rosenthal: danielkitairosenthal@gmail.com
David Rosenthal: rosenthd@stjohns.edu
Peter Rosenthal: rosent@math.toronto.edu

Toronto, ON, Canada   Daniel Rosenthal
Queens, NY, USA   David Rosenthal
Toronto, ON, Canada   Peter Rosenthal

Contents

Chapter 1
Introduction to the Natural Numbers

We assume basic knowledge about the numbers that we count with; that is, the numbers 1, 2, 3, 4, 5, 6, and so on. These are called the *natural numbers*, and the set consisting of all of them is usually denoted by $\mathbb{N}$. They do seem to be very natural, in the sense that they arose very early on in virtually all societies. There are many other names for these numbers, such as the *positive integers* and the *positive whole numbers*. Although the natural numbers are very familiar, we will see that they have many interesting properties beyond the obvious ones. Moreover, there are many questions about the natural numbers to which nobody knows the answer. Some of these questions can be stated very simply, as we shall see, although their solution has eluded the thousands of mathematicians who have attempted to solve them.

We assume familiarity with the two basic operations on the natural numbers, addition and multiplication. The sum of two numbers will be indicated using the plus sign "+." Multiplication will be indicated by putting a dot in the middle of the line between the numbers, or by simply writing the symbols for the numbers next to each other, or sometimes by enclosing them in parentheses. For example, the product of 3 and 2 could be denoted $3 \cdot 2$ or $(3)(2)$. The product of the natural numbers represented by the symbols m and n could be denoted mn, or $m \cdot n$, or $(m)(n)$.

We also, of course, need the number 0. Moreover, we require the negative whole numbers as well. For each natural number n there is a corresponding negative number $-n$ such that $n + (-n) = 0$. Altogether, the collection of positive and negative numbers and 0 is called the *integers*. It is often denoted by $\mathbb{Z}$.

We assume that you know how to add two negative integers and also how to add a negative integer to a positive integer. Multiplication appears to be a bit more mysterious. Most people feel comfortable with the fact that, for m and n natural numbers, the product of m and $(-n)$ is $-mn$. What some people find more mysterious is the fact that $(-m)(-n) = mn$ for natural numbers m and n; that is, the product of two negative integers is a positive integer. There are various possible explanations that can be provided for this, one of which is the following. Using the usual rules of arithmetic:

D. Rosenthal et al., *A Readable Introduction to Real Mathematics*,
Undergraduate Texts in Mathematics, DOI 10.1007/978-3-319-05654-8_1,
© Springer International Publishing Switzerland 2014

$$(-m)(-n) + (-m)(n) = (-m)(-n + n) = (-m)(0) = 0$$

Adding mn to both sides of this equation gives

$$(-m)(-n) + (-m)(n) + mn = 0 + mn$$

or

$$(-m)(-n) + \big((-m) + m\big) \cdot n = mn$$

Thus,

$$(-m)(-n) + 0 \cdot n = mn$$

so,

$$(-m)(-n) = mn$$

Therefore, the fact that $(-m)(-n) = mn$ is implied by the other standard rules of arithmetic.

1.1 Prime Numbers

One of the important concepts we will study is *divisibility*. For example, 12 is divisible by 3, which means that there is a natural number (in this case, 4) such that the product of 3 and that natural number is 12. That is, $12 = 3 \cdot 4$. In general, we say that *the integer m is divisible by the integer n* if there is an integer q such that $m = nq$. There are many other terms that are used to describe such a relationship. For example, if $m = nq$, we may say that n and q are *divisors* of m and that each of n and q *divides m*. The terminology "q is the quotient when m is divided by n" is also used when n is different from 0. In this situation, n and q are also sometimes called *factors* of m; the process of writing an integer as a product of two or more integers is called *factoring* the integer.

The number 1 is a divisor of every natural number since, for each natural number m, $m = 1 \cdot m$. Also, every natural number m is a divisor of itself, since $m = m \cdot 1$.

The number 1 is the only natural number that has only one natural number divisor, namely itself. Every other natural number has at least two divisors, itself and 1. The natural numbers that have exactly two natural number divisors are called *prime numbers*. That is, a *prime number* is a natural number greater than 1 whose only natural number divisors are 1 and the number itself. We do not consider the number 1 to be a prime; the first prime number is 2. The primes continue: 3, 5, 7, 11, 13, 17, 19, 23, 29, 31, and so on.

And so on? Is there a largest prime? Or does the sequence of primes continue without end? There is, of course, no largest natural number. For if n is any natural number, then $n + 1$ is a natural number and $n + 1$ is bigger than n. It is not so easy to determine if there is a largest prime number or not. If p is a prime, then $p + 1$ is almost never a prime. Of course, if $p = 2$, then $p + 1 = 3$ and p and $p + 1$ are both primes. However, 2 is the only prime number p for which $p + 1$ is prime. This can be proven as follows. First note that, since every even number is divisible by 2, 2 itself is the only even prime number. Therefore, if p is a prime other than 2, then p is odd and $p + 1$ is an even number larger than 2 and is thus not prime.

Is it nonetheless true that, given any prime number p, there is a prime number larger than p? Although we cannot get a larger prime by simply adding 1 to a given prime, there may be some other way of producing a prime larger than any given one. We will answer this question after learning a little more about primes.

A natural number, other than 1, that is not prime is said to be *composite*. (The number 1 is special and is neither prime nor composite.) For example, 4, 68, 129, and 2010 are composites. Thus, a composite number is a natural number other than 1 that has a divisor in addition to itself and 1.

To determine if a number is prime, what potential factors must be checked to eliminate the possibility that there are factors other than the number and 1? If $m = n \cdot q$, it is not possible that n and q are both larger than the square root of m, for if two natural numbers are both larger than the square root of m, then their product is larger than m. It follows that a natural number (other than 1) that is not prime has at least one divisor that is larger than 1 and is no larger than the square root of that natural number. Thus, to check whether or not a natural number m is prime, you need not check whether every natural number less than m divides m. It suffices to check if m has a divisor that is larger than 1 and no larger than the square root of m. If it has such a divisor, it is composite; if it has no such divisor, it is prime.

For example, we can conclude that 101 is prime since none of the numbers 2, 3, 4, 5, 6, 7, 8, 9, 10 are divisors of 101.

Using refinements of this idea and powerful computers, many very large numbers have been shown to be prime. For example, 100,000,559 is prime, as is 22,801,763,489.

The fact that very large natural numbers have been shown to be prime does not answer the question of whether there is a largest prime. The theorem that there is always a prime larger than p for *every* prime number p cannot be established by computing any number of specific primes, no matter how large.

Over the centuries, mathematicians have discovered many proofs that there is no largest prime. We shall present one of the simplest and most beautiful proofs, discovered by the Ancient Greeks.

We begin by establishing a preliminary fact that is required for the proof. A statement that is proven for the purpose of being used to prove something else is called a "lemma." We need a lemma. The lemma that we require states that every composite number has a divisor that is a prime number. (The proof that we present of the lemma is quite convincing, but we shall subsequently present a more precise proof.)

Lemma 1.1.1. *Every natural number greater than 1 has a prime divisor.*

Proof. If the number is prime, it is a divisor of itself. If the number, say m, is composite, then m has at least one factorization $m = n \cdot q$, where neither n nor q is m or 1. If either of n or q is a prime number, then the lemma is established for m. If n is not prime, then it has a factorization $n = s \cdot t$, where s and t are natural numbers other than 1 and n. It is clear that s and t are also divisors of m. Thus, if either of s and t is a prime number, the lemma is established. If s is not prime, then it can be factored into a product where neither factor is s or 1 and so on. Continued factoring must get down to a factor that cannot itself be factored, i.e., to a prime. That prime number is a divisor of m, so the lemma is established.   □

The following is the ingenious proof of the infinitude of the primes discovered by the Ancient Greeks.

Theorem 1.1.2. *There is no largest prime number.*

Proof. Let p be any prime number. We must prove that there is some prime larger than p. To do this, we will construct a number that we will show is either a prime larger than p or has a prime divisor larger than p. In both cases we will conclude that there is a prime number larger than p.

Here is how we construct the large number. Let M be the number obtained by taking the product of all the prime numbers up to and including the given prime p and then adding 1 to that product. That is,

$$M = (2 \cdot 3 \cdot 5 \cdot 7 \cdot 11 \cdot 13 \cdot 17 \cdot 19 \cdots p) + 1$$

It is possible that M is a prime number. If that is so, then there is a prime number larger than p, since M is obviously larger than p. If M is not prime, then it is composite. We must show that there is a prime larger than p in this case as well.

Suppose, then, that M is composite. By Lemma 1.1.1, it follows that M has a prime divisor. Let q be any prime divisor of M. We will show that q is larger than p and thus that there is a prime larger than p in this case as well.

Consider possible values of q, a prime divisor of M. Surely q is not 2, for

$$2 \cdot 3 \cdot 5 \cdot 7 \cdot 11 \cdot 13 \cdot 17 \cdot 19 \cdots p$$

is an even number, and thus adding 1 to that number to get M produces an odd number. That is, M is odd and is therefore not divisible by 2. Since q does divide M, q cannot be equal to 2.

Similar reasoning shows that q cannot be 3. For

$$2 \cdot 3 \cdot 5 \cdot 7 \cdot 11 \cdot 13 \cdot 17 \cdot 19 \cdots p$$

is a multiple of 3, so the number obtained by adding 1, namely M, leaves a remainder of 1 when it is divided by 3. That is, 3 is not a divisor of M. Since q is a divisor of M, q is not 3.

Exactly the same proof shows that q is not 5, since 5 is a divisor of

$$2 \cdot 3 \cdot 5 \cdot 7 \cdot 11 \cdot 13 \cdot 17 \cdot 19 \cdots p$$

and thus cannot be a divisor of M. In fact, the same proof establishes that q cannot be any of the factors $2, 3, 5, \ldots, p$ of the product

$$2 \cdot 3 \cdot 5 \cdot 7 \cdot 11 \cdot 13 \cdot 17 \cdot 19 \cdots p$$

Since every prime number up to and including p is a factor of that product, q cannot be any of those prime numbers. Therefore q is a prime number that is not any of the prime numbers up to and including p. It follows that q is a prime number larger than p, and we have proven that there is a prime number larger than p in the case where M is composite. Therefore, in both cases, the case where M is prime and the case where M is composite, we have shown that there is a prime number larger than p. This proves the theorem.   □

Every mathematician would agree that the above proof is "elegant." If you find the proof interesting, then you are likely to appreciate many of the other ideas that we will discuss (and much mathematics that we do not cover as well).

1.2   Unanswered Questions

There are many questions concerning prime numbers that no one has been able to answer. One famous question concerns what are called *twin primes*. Since 2 is the only even prime number, the only consecutive integers that are prime are 2 and 3. There are, however, many pairs of primes that are two apart, such as $\{3, 5\}$, $\{29, 31\}$, $\{101, 103\}$, $\{1931, 1933\}$, and $\{104471, 104473\}$. Such pairs are called *twin primes*. One question that remains unanswered, in spite of the efforts of thousands of mathematicians over hundreds of years, is the question of whether there is a largest pair of twin primes. Some very large pairs are known (e.g., $\{1000000007, 1000000009\}$ and many pairs that are even much bigger), but no one knows if there is a largest such.

Another very famous unsolved problem is whether or not the *Goldbach Conjecture* is true. Several hundred years ago, Goldbach Conjectured (that is, said that he thought that it was probably true) that every even natural number larger than 2 is the sum of two prime numbers (e.g., $6 = 3 + 3$, $20 = 7 + 13$, and $22{,}901{,}764{,}048 = 22{,}801{,}763{,}489 + 100{,}000{,}559$). Goldbach's Conjecture is known to be true for many very large even natural numbers, but no one has been able to prove it in general (or to show that there is an even number that cannot be written as the sum of two primes).

If you are able to solve the Twin Primes Problem or determine the truth or falsity of Goldbach's Conjecture, you will immediately become famous throughout the world and your name will remain famous as long as civilization endures. On the other hand, it will almost undoubtedly prove to be extremely difficult to answer either of those questions. On the other "other hand," there is a very slight possibility that one or both of those questions have a fairly simple answer that has been overlooked by the many great and not-so-great mathematicians who have thought about them. In spite of the small possibility of success you might find it interesting to think about these problems.

1.3   Problems

Basic Exercises

1. Show that the following are composite numbers:

    (a) 68
    (b) 129
    (c) 20,101,116

2. Which of the following are prime numbers?

    (a) 79
    (b) 153
    (c) 537
    (d) 851,486

3. Write each of the following numbers as a sum of two primes.

    (a) 100
    (b) 112

Interesting Problems

4. Verify that the Goldbach Conjecture holds for all even numbers up to 100.
5. Find a pair of twin primes such that each prime is greater than 1000.

Challenging Problems

6. Find a prime number p such that the number $(2 \cdot 3 \cdot 5 \cdot 7 \cdots p) + 1$ is not prime.
7. Suppose that p, $p + 2$, and $p + 4$ are prime numbers. Prove that $p = 3$.
    [Hint: Why can't p be 5 or 7?]

8. Prove that, for every natural number $n > 2$, there is a prime number between n and $n!$. (Recall that $n!$ is defined to be $n(n-1)(n-2)\cdots 2 \cdot 1$.)
   [Hint: There is a prime number that divides $n! - 1$.]
   Note that this gives an alternate proof that there are infinitely many prime numbers.

9. Prove that, for every natural number n, there are n consecutive composite numbers.
   [Hint: $(n+1)! + 2$ is a composite number.]

10. Show that a natural number has an odd number of different factors if and only if it is a perfect square (i.e., it is the square of another natural number).

Chapter 2
Mathematical Induction

There is a method for proving certain theorems that is called *mathematical induction*. We will give a number of examples of proofs that use this method. The basis for mathematical induction, however, is a statement about sets of natural numbers. Recall that the set of all natural numbers is the set $\{1, 2, 3, \ldots\}$. Mathematical induction provides an alternate description of that set.

2.1 The Principle of Mathematical Induction

Suppose S is a set of natural numbers that has the following two properties:

A. *The number 1 is in S.*
B. *Whenever a natural number is in S, the next natural number is also in S.*

The second property can be stated a little more formally: If k is a natural number and k is in S, then $k + 1$ is in S.

What can we say about a set S that has those two properties? Since 1 is in S (by property A), it follows from property B that 2 is in S. Since 2 is in S, it follows from property B that 3 is in S. Since 3 is in S, 4 is in S. Then 5 is in S, 6 is in S, 7 is in S, and so on. It seems clear that S must contain every natural number. That is, the only set of natural numbers with the above two properties is the set of all natural numbers. We state this formally:

The Principle of Mathematical Induction 2.1.1. *If S is any set of natural numbers with the properties that*

A. *1 is in S, and*
B. *$k + 1$ is in S whenever k is any number in S,*

then S is the set of all natural numbers.

D. Rosenthal et al., *A Readable Introduction to Real Mathematics*,
Undergraduate Texts in Mathematics, DOI 10.1007/978-3-319-05654-8_2,
© Springer International Publishing Switzerland 2014

We gave an indication above of why the Principle of Mathematical Induction is true. A more formal proof can be based on the following more obvious fact, which we assume as an axiom.

The Well-Ordering Principle 2.1.2. *Every set of natural numbers that contains at least one element has a smallest element in it.*

We can establish the Principle of Mathematical Induction from the Well-Ordering Principle as follows. Suppose that the Well-Ordering Principle holds for all sets of natural numbers. Let S be any set of natural numbers and suppose that S has properties A and B of the Principle of Mathematical Induction. To prove the Principle of Mathematical Induction, we must prove that the only such set S is the set of all natural numbers. We will do this by showing that it is impossible that there is any natural number that is not in S. To see this, suppose that S does not contain all natural numbers. Then let T denote the set of all natural numbers that are not in S. Assuming that S is not the set of all natural numbers is equivalent to assuming that T has at least one element. If this were the case, then well-ordering would imply that T has a smallest element. We will show that this is impossible.

Suppose that t was the smallest element of T. Since 1 is in S, 1 is not in T. Therefore, t is larger than 1, so $t - 1$ is a natural number. Since $t - 1$ is less than the smallest number t in T, $t - 1$ cannot be in T. Since T contains all the natural numbers that are not in S, it follows that $t - 1$ is in S. This, however, leads to the following contradiction. Since S has property B, $(t - 1) + 1$ must also be in S. But this is t, which is in T and therefore not in S. This shows that the assumption that there is a smallest element of T is not consistent with the properties of S. Thus, there is no smallest element of T and, by well-ordering, there is therefore no element in T. This proves that S is the set of all natural numbers.

The way mathematical induction is usually explained can be illustrated by considering the following example. Suppose that we wish to prove, for every natural number n, the validity of the following formula for the sum of the first n natural numbers:

$$1 + 2 + 3 + \cdots + (n - 1) + n = \frac{n(n + 1)}{2}$$

One way to prove that this formula holds for every n is the following. First, the formula does hold for $n = 1$, for in this case the left-hand side is just 1 and the right-hand side is $\frac{1 \cdot (1+1)}{2}$, which is equal to 1. To prove that the formula holds for all n, we will establish the fact that whenever the formula holds for any given natural number, the formula will also hold for the next natural number. That is, we will prove that the formula holds for $n = k + 1$ whenever it holds for $n = k$. (This passage from k to $k + 1$ is often called "the inductive step.") If we prove this fact, then, since we know that the formula does hold for $n = 1$, it would follow from this fact that it holds for the next natural number, 2. Then, since it holds for $n = 2$, it holds for the natural number that follows 2, which is 3. Since it holds for 3, it holds for 4, and then for 5, and 6, and so on. Thus, we will conclude that the formula holds for every natural number. (This is really just the Principle of Mathematical Induction as we formally stated it above. If S is the set of all n for which the formula for the sum is true, showing that S has properties A and B leads to the conclusion that S is the set of all natural numbers.)

To prove the formula in general, then, we must show that the formula holds for $n = k + 1$ whenever it holds for $n = k$. Assume that the formula does hold for $n = k$, where k is any fixed natural number. That is, we assume the formula

$$1 + 2 + 3 + \cdots + (k - 1) + k = \frac{k(k + 1)}{2}$$

We want to derive the formula for $n = k + 1$ from the above equation. That is easy to do, as follows. Assuming the above formula, add $k + 1$ to both sides, getting

$$1 + 2 + 3 + \cdots + (k - 1) + k + (k + 1) = \frac{k(k + 1)}{2} + (k + 1)$$

We shall see that a little algebraic manipulation of the right-hand side of the above will produce the formula for $n = k + 1$. To see this, simply note that

$$\frac{k(k + 1)}{2} + (k + 1) = \frac{k(k + 1)}{2} + \frac{2(k + 1)}{2}$$
$$= \frac{k(k + 1) + 2(k + 1)}{2}$$
$$= \frac{(k + 2)(k + 1)}{2}$$
$$= \frac{(k + 1)(k + 2)}{2}$$
$$= \frac{(k + 1)((k + 1) + 1)}{2}$$

Thus,

$$1 + 2 + 3 + \cdots + (k - 1) + k + (k + 1) = \frac{(k + 1)((k + 1) + 1)}{2}$$

This equation is the same as that obtained from the formula by substituting $k + 1$ for n. Therefore we have established the inductive step, so we conclude that the formula does hold for all n.

There are many very similar proofs of similar formulas.

Theorem 2.1.3. *For every natural number n,*

$$1^2 + 2^2 + 3^2 + \cdots + n^2 = \frac{n(n + 1)(2n + 1)}{6}$$

Proof. Let S be the set of all natural numbers for which the theorem is true. We want to show that S contains all of the natural numbers. We do this by showing that S has properties A and B.

For property A, we need to check that $1^2 = \frac{1(1+1)(2\cdot1+1)}{6}$. This is true, so S satisfies property A. To verify property B, let k be in S. We must show that $k+1$ is in S. Since k is in S, the theorem holds for k. That is,

$$1^2 + 2^2 + 3^2 + \cdots + k^2 = \frac{k(k+1)(2k+1)}{6}$$

Using this formula, we can prove the corresponding formula for $k+1$ as follows. Adding $(k+1)^2$ to both sides of the above equation, we get

$$1^2 + 2^2 + 3^2 + \cdots + k^2 + (k+1)^2 = \frac{k(k+1)(2k+1)}{6} + (k+1)^2$$

Now we do some algebraic manipulations to the right-hand side to see that it is what we want:

$$\frac{k(k+1)(2k+1)}{6} + (k+1)^2 = \frac{k(k+1)(2k+1) + 6(k+1)^2}{6}$$

$$= \frac{(k+1)\Big(k(2k+1) + 6(k+1)\Big)}{6}$$

$$= \frac{(k+1)\Big((2k^2+k) + (6k+6)\Big)}{6}$$

$$= \frac{(k+1)(2k^2 + 7k + 6)}{6}$$

$$= \frac{(k+1)(k+2)(2k+3)}{6}$$

The last equation is the formula in the case when $n = k+1$, so $k+1$ is in S. Therefore, S is the set of natural numbers by the Principle of Mathematical Induction. $\square$

Sometimes one wants to prove something by induction that is not true for all natural numbers, but only for those bigger than a given natural number. A slightly more general principle that we can use in such situations is the following.

The Generalized Principle of Mathematical Induction 2.1.4. *Let m be a natural number. If S is a set of natural numbers with the properties that*

A. m is in S, and
B. k + 1 is in S whenever k is in S and is greater than or equal to m,

then S contains every natural number greater than or equal to m.

The Principle of Mathematical Induction is the special case of the generalized principle when $m = 1$. The generalized principle states that we can use induction starting at any natural number, not just at 1.

For example, consider the question: which is larger, $n!$ or 2^n? Recall that $n! = n \cdot (n-1) \cdot (n-2) \cdots 3 \cdot 2 \cdot 1$. For $n = 1, 2$, and 3, we see that

$$1! = 1 < 2^1 = 2$$

$$2! = 2 \cdot 1 = 2 < 2^2 = 2 \cdot 2 = 4$$

$$3! = 3 \cdot 2 \cdot 1 = 6 < 2^3 = 2 \cdot 2 \cdot 2 = 8$$

But when $n = 4$, the inequality is reversed, since

$$4! = 4 \cdot 3 \cdot 2 \cdot 1 = 24 > 2^4 = 2 \cdot 2 \cdot 2 \cdot 2 = 16$$

When $n = 5$,

$$5! = 5 \cdot 4 \cdot 3 \cdot 2 \cdot 1 = 120 > 2^5 = 2 \cdot 2 \cdot 2 \cdot 2 \cdot 2 = 32$$

If you think about it a bit, it is clear why eventually $n!$ is much bigger than 2^n. In both expressions we are multiplying n numbers together, but for 2^n we are always multiplying by 2, whereas the numbers we multiply to build $n!$ get larger and larger. While it is not true that $n! > 2^n$ for every natural number (since it is not true for $n = 1, 2$, and 3), we can, as we now show, use the more general form of mathematical induction to prove that it is true for all natural numbers greater than or equal to 4.

Theorem 2.1.5. $n! > 2^n$ *for* $n \geq 4$.

Proof. We use the Generalized Principle of Mathematical Induction with $m = 4$. Let S be the set of natural numbers for which the theorem is true. As we saw above, $4! > 2^4$. Therefore, 4 is in S. Thus, property A is satisfied. For property B, assume that $k \geq 4$ and that k is in S; i.e., $k! > 2^k$. We must show that $(k+1)! > 2^{k+1}$. Multiplying both sides of the inequality for k (which we have assumed to be true) by $k + 1$ gives

$$(k+1)(k!) > (k+1) \cdot 2^k$$

The left-hand side is just $(k+1)!$; therefore we have the inequality

$$(k+1)! > (k+1) \cdot 2^k$$

Since $k \geq 4$, $k+1 > 2$. Therefore, the right-hand side of the inequality, $(k+1) \cdot 2^k$, is greater than $2 \cdot 2^k = 2^{k+1}$. Combining this with the above inequality, we get

$$(k+1)! > (k+1) \cdot 2^k > 2^{k+1}$$

Thus, $k + 1$ is in S, which verifies property B. By the Generalized Principle of Mathematical Induction, S contains all natural numbers greater than or equal to 4.

□

The following is an example where mathematical induction is useful in establishing a geometric result. We will use the word "tromino" to denote an L-shaped object consisting of three squares of the same size. That is, a tromino looks like this:

Another way to think of a tromino is that it is the geometric figure obtained by taking a square that is composed of four smaller squares and removing one of the smaller squares.

We are going to consider what geometric regions can be covered by trominos, all of which have the same size and that do not overlap each other. As a first example, start with a square made up of 16 smaller squares (i.e., a square that is "4 by 4") and remove one small square from a corner of the square:

Can the region that is left be covered by trominos (each made up of three small squares of the same size as the small squares in the region) that do not overlap each other? It can:

We can use mathematical induction to prove the following.

Theorem 2.1.6. *For each natural number n, consider a square consisting of 2^{2n} smaller squares. (That is, a $2^n \times 2^n$ square.) If one of the smaller squares is removed from a corner of the large square, then the resulting region can be completely covered by trominos (each made up of three small squares of the same size as the small squares in the region) in such a way that the trominos do not overlap.*

Proof. To begin a proof by mathematical induction, first note that the theorem is certainly true for $n = 1$; the region obtained after removing a small corner square is a tromino, so it can be covered by one tromino.

Suppose that the theorem is true for $n = k$. That is, we are supposing that if a small corner square is removed from any $2^k \times 2^k$ square consisting of 2^{2k} smaller squares, then the resulting region can be covered by trominos. The proof will be established by the Principle of Mathematical Induction if we can show that the same result holds for $n = k + 1$. Consider, then, any $2^{k+1} \times 2^{k+1}$ square consisting of smaller squares. Remove one corner square to get a region that looks like this:

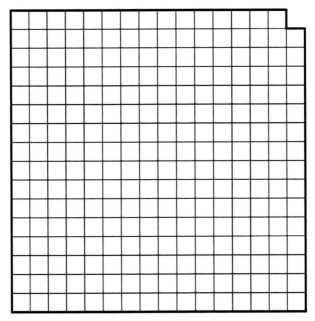

The region can be divided into four "medium-sized" squares that are each $2^k \times 2^k$, like this:

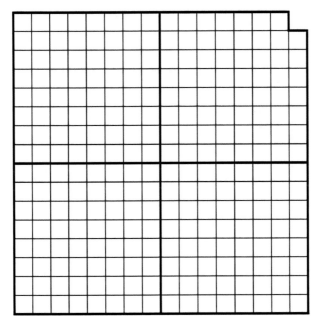

Now place a tromino in the middle of the region, as illustrated below.

The four "medium-sized" squares of the region are each $2^k \times 2^k$ and, because of the tromino in the middle, the "medium-sized" squares remaining to be covered each have one corner covered or missing.

By the inductive hypothesis, trominos can be used to cover the rest of each of the four "medium-sized" squares. This leads to a covering of the entire $2^{k+1} \times 2^{k+1}$ square, thus finishing the proof by mathematical induction.   □

2.2 The Principle of Complete Mathematical Induction

There is a variant of the Principle of Mathematical Induction that is sometimes very useful. The basis for this variant is a slightly different characterization of the set of all natural numbers.

The Principle of Complete Mathematical Induction 2.2.1. *(Sometimes called "the Principle of Strong Mathematical Induction.") If S is any set of natural numbers with the properties that*

A. *1 is in S, and*
B. *$k + 1$ is in S whenever k is a natural number and all of the natural numbers from 1 through k are in S,*

then S is the set of all natural numbers.

The informal and formal proofs of the Principle of Complete Mathematical Induction are virtually the same as the proofs of the Principle of (ordinary) Mathematical Induction. First consider the informal proof. If S is any set of natural numbers with properties A and B of the Principle of Complete Mathematical Induction, then, in particular, 1 is in S. Since 1 is in S, it follows from property B that 2 is in S. Since 1 and 2 are in S, it follows from property B that 3 is in S. Since 1, 2, and 3 are in S, 4 is in S and so on. It is suggested that you write out the details of the formal proof of the Principle of Complete Mathematical Induction as a consequence of the Well-Ordering Principle.

Just as for ordinary induction, the Principle of Complete Mathematical Induction can be generalized to begin at any natural number, not just 1.

The Generalized Principle of Complete Mathematical Induction 2.2.2. *If S is any set of natural numbers with the properties that*

A. *m is in S, and*
B. *$k + 1$ is in S whenever k is a natural number greater than or equal to m and all of the natural numbers from m through k are in S,*

then S contains all natural numbers greater than or equal to m.

There are many situations in which it is difficult to directly apply the Principle of Mathematical Induction but easy to apply the Principle of Complete Mathematical Induction. One example of such a situation is a very precise proof of the lemma (Lemma 1.1.1) that was required to prove that there is no largest prime number.

Lemma 2.2.3. *Every natural number greater than 1 has a prime divisor.*

The following is a statement that clearly implies the above lemma. Note that we employ the convention that a single prime number is a "product of primes" where the product has only one factor.

Theorem 2.2.4. *Every natural number other than 1 is a product of prime numbers.*

Proof. We prove this theorem using the Generalized Principle of Complete Mathematical Induction starting at 2. Let S be the set of all n that are products of primes. It is clear that 2 is in S, since 2 is a prime. Suppose that every natural number from 2 up to k is in S. We must show, in order to apply the Generalized Principle of Complete Mathematical Induction, that $k + 1$ is in S.

The number $k + 1$ cannot be 1. We must therefore show that either it is prime or is a product of primes. If $k + 1$ is prime, we are done. If $k + 1$ is not prime, then $k + 1 = xy$ where each of x and y is a natural number strictly between 1 and $k + 1$. Thus x and y are each at most k, so, by the inductive hypothesis, x and y are both in S. That is, x and y are each either primes or the product of primes. Therefore, xy can be written as a product of primes by writing the product of the primes comprising x (or x itself if x is prime) times the product of the primes comprising y (or y itself if y is prime). Thus, by the Generalized Principle of Complete Mathematical Induction starting at 2, S contains all natural numbers greater than or equal to 2.   □

We now describe an interesting theorem whose statement is a little more difficult to understand. (If you find this theorem too difficult, you need not consider it; it won't be used in anything that follows. You might wish to return to it at some later time.)

We begin by describing the case where $n = 5$. Suppose there is a pile of 5 stones. We are going to consider the sum of certain sequences of numbers obtained as follows. Begin one such sequence by dividing the pile into two smaller piles, a pile of 3 stones and a pile of 2 stones. Let the first term in the sum be $3 \cdot 2 = 6$. Repeat this process with the pile of 3 stones: divide it into a pile of 2 stones and a pile consisting of 1 stone. Add $2 \cdot 1 = 2$ to the sum. The pile with 2 stones can be divided into 2 piles of 1 stone each. Add $1 \cdot 1 = 1$ to the sum. Now go back to the pile of 2 stones created by the first division. That pile can be divided into 2 piles of 1 stone each. Add $1 \cdot 1 = 1$ to the sum. The total sum that we have is 10.

Let's create another sum in a similar manner but starting a different way. Divide the original pile of 5 stones into a pile of 4 stones and a pile of 1 stone. Begin this sum with $4 \cdot 1 = 4$. Divide the pile of 4 stones into two piles of 2 stones each and add $2 \cdot 2 = 4$ to the sum. The first pile of 2 stones can be divided into two piles of 1 stone each, so add $1 \cdot 1 = 1$ to the sum. Similarly, divide the second pile of 2 into two piles of 1 each and add $1 \cdot 1 = 1$ to the sum. The sum we get proceeding in this way is also 10.

Is it a coincidence that we got the same result, 10, for the sums we obtained in quite different ways?

Theorem 2.2.5. *For any natural number n greater than 1, consider a pile of n stones. Create a sum as follows: divide the given pile of stones into two smaller piles. Let the product of the number of elements in one smaller pile and the number of elements in the other smaller pile be the first term in the sum. Then consider one of the smaller piles and (unless it consists of only one stone) divide that pile into two smaller piles and let the product of the number of stones in those piles be the second term in the sum. Do the same for the other smaller pile. Continue dividing, multiplying, and adding terms to the sum in all possible ways. No matter what sequence of divisions into subpiles is used, the total sum is $n(n-1)/2$.*

Proof. We prove this theorem using Generalized Complete Mathematical Induction beginning with $n = 2$. Given any pile of 2 stones, there is only one way to divide it: into two piles of 1 each. Since $1 \cdot 1 = 1$, the sum is 1 in this case. Notice that $1 = 2(2-1)/2$, so the formula holds for the case $n = 2$.

Suppose now that the formula holds for all of $n = 2, 3, 4, \ldots, k$. Consider any pile of $k+1$ stones. Note that $k+1$ is at least 3. We must show that for any sequence of divisions, the resulting sum is $(k+1)(k+1-1)/2 = k(k+1)/2$.

Begin with any division of the pile into two subpiles. Call the number of stones in the subpiles x and y respectively. Consider first the situation where $x = 1$. Then the first term in the sum is $1 \cdot y = y$. Since $x = 1$ and $x + y = k+1$, we know that $y = k$. The process is continued by dividing the pile of y stones. By the inductive hypothesis (since $y = k$, which is greater than or equal to 2), the sum obtained by completing the process on a pile of y stones is $y(y-1)/2$. Thus, the total sum for the original pile of $k+1$ stones in this case is

$$y + \frac{y(y-1)}{2} = \frac{2y + (y^2 - y)}{2} = \frac{y^2 + y}{2} = \frac{y(y+1)}{2} = \frac{k(k+1)}{2}$$

If $y = 1$, the same proof can be given by simply interchanging the roles of x and y in the previous paragraph.

The last, and most interesting, case is when neither x nor y is 1. In this case, both x and y are greater than or equal to 2 and less than k. The first term in the sum is then xy. Continuing the process will give a total sum that is equal to xy plus the sum for the pile of x stones added to the sum for the pile of y stones. Therefore, using the inductive hypothesis, the sum for the original pile of $k+1$ stones is $xy + x(x-1)/2 + y(y-1)/2$. We must show that this sum is $k(k+1)/2$.

Recall that $k + 1 = x + y$, so $x = k + 1 - y$. Using this, we see that

$$xy + \frac{x(x-1)}{2} + \frac{y(y-1)}{2}$$
$$= \frac{2(k+1-y)y}{2} + \frac{(k+1-y)(k-y)}{2} + \frac{y(y-1)}{2}$$
$$= \frac{2ky + 2y - 2y^2}{2} + \frac{k^2 + k - ky - ky - y + y^2}{2} + \frac{y^2 - y}{2}$$
$$= \frac{k^2 + k}{2}$$
$$= \frac{k(k+1)}{2}$$

This completes the proof.   □

Mathematics is the most precise of subjects. However, human beings are not always so precise; they must be careful not to make mistakes. See if you can figure out what is wrong with the "proof" of the following obviously false statement.

False Statement. All human beings are the same age.

"Proof". We will present what, at first glance at least, appears to be a proof of the above statement. We begin by reformulating it as follows: For every natural number n, every set of n people consists of people the same age. The assertion that "all human beings are the same age" would clearly follow as the case where n is the present population of the earth. We proceed by mathematical induction. The case $n = 1$ is certainly true; a set containing 1 person consists of people the same age. For the inductive step, suppose that every set of k people consists of people the same age. Let S be any set containing $k + 1$ people. We must show that all the people in S are the same age as each other.

List the people in S as follows:

$$S = \{P_1, P_2, \ldots, P_k, P_{k+1}\}$$

Consider the subset L of S consisting of the first k people in S; that is,

$$L = \{P_1, P_2, \ldots, P_k\}$$

Similarly, let R denote the subset consisting of the last k elements of S; that is,

$$R = \{P_2, \ldots, P_k, P_{k+1}\}$$

The sets L and R each contain k people, and so by the inductive hypothesis each consists of people who are the same age as each other. In particular, all the people in L are the same age as P_2. Also, all the people in R are the same age as P_2. But every person in the original set S is in either L or R, so all the people in S are the same age as P_2. Therefore, S consists of people the same age, and the assertion follows by the Principle of Mathematical Induction.

What is going on? Is it really true that all people are the same age? Not likely. Is the Principle of Mathematical Induction flawed? Or is there something wrong with the above "proof"?

Clearly there must be something wrong with the "proof." Please do not read further for at least a few minutes while you try to find the mistake.

Wait a minute. Before you read further, please try for a little bit longer to see if you can find the mistake.

If you haven't been able to find the error yourself, perhaps a hint will help. The proof of the case $n = 1$ is surely valid; a set with one person in it contains a person with whatever age that person is. What about the inductive step, going from k to $k + 1$? For it to be valid, it must apply for every natural number k. To conclude that an assertion holds for all natural numbers given that it holds for $n = 1$ requires that its truth for $n = k + 1$ is implied by its truth for $n = k$, *for every natural number k.* In fact, there is a k for which the above derivation of the case $n = k + 1$ from the case $n = k$ is not valid. Can you figure out the value of that k?

Okay, here is the mistake. Consider the inductive step when $k = 1$; that is, going from 1 to 2. In this case, the set S would have the form

$$S = \{P_1, P_2\}$$

Then, $L = \{P_1\}$ and $R = \{P_2\}$.

The set L does consist of people the same age as each other, as does the set R. But there is no person who is in both sets. Thus, we cannot conclude that everyone in S is the same age. This shows that the above "proof" of the inductive step does not hold when $k = 1$. In fact, the following is true.

True Statement. If every pair of people in a given set of people consists of people the same age, then all the people in the set are the same age.

Proof. Let S be the given set of people; suppose $S = \{P_1, P_2, \ldots, P_n\}$. For each i from 2 to n, the pair $\{P_1, P_i\}$ consists of people the same age, by hypothesis. Thus, P_i and P_1 are the same age for every i, so every person in S is the same age as P_1. Hence, everyone in S is the same age.   □

2.3   Problems

Basic Exercises

1. Prove, using induction, that for every natural number n:

$$1 \cdot 2 + 2 \cdot 3 + 3 \cdot 4 + \cdots + n \cdot (n+1) = \frac{n(n+1)(n+2)}{3}$$

2. Prove, using induction, that for every natural number n:

$$\frac{1}{1 \cdot 2} + \frac{1}{2 \cdot 3} + \cdots + \frac{1}{n \cdot (n+1)} = \frac{n}{n+1}$$

3. Prove, using induction, that for every natural number n:

$$2 + 2^2 + 2^3 + \cdots + 2^n = 2^{n+1} - 2$$

4. Prove, using induction, that for every natural number n:

$$\frac{1}{2} + \frac{2}{2^2} + \frac{3}{2^3} + \cdots + \frac{n}{2^n} = 2 - \frac{n+2}{2^n}$$

Interesting Problems

5. Prove the following statement by induction: For every natural number n, every set with n elements has 2^n subsets. (Note that the empty set is a subset of every set.)

6. Prove, using induction, that for every natural number n:

$$1 + \frac{1}{\sqrt{2}} + \frac{1}{\sqrt{3}} + \cdots + \frac{1}{\sqrt{n}} < 2\sqrt{n}$$

7. Prove by induction that 3 divides $n^3 + 2n$, for every natural number n.

8. Show that $3^n > n^2$ for every natural number n.

9. Use induction to prove that $2^n > n^2$, for every $n > 4$.

10. Show that for every natural number $n > 1$ and every real number r different from 1:

$$1 + r + r^2 + \cdots + r^{n-1} = \frac{r^n - 1}{r - 1}$$

Challenging Problems

11. Prove the Principle of Complete Mathematical Induction using the Well-Ordering Principle.
12. Prove the Well-Ordering Principle using the Principle of Complete Mathematical Induction.
13. One version of a game called *Nim* is played as follows. There are two players and two piles consisting of the same natural number of objects; for this example, suppose the objects are nickels. At each turn, a player removes some number of nickels from either one of the piles. Then the other player removes some number of nickels from either of the piles. The players continue playing alternately until the last nickel is removed. The winner is the player who removes the last nickel.

    Prove: If the second player always removes the same number of nickels that the first player last removed and does so from the other pile (thus making the piles equal in number after the second player's turn), then the second player will win.
14. Define the *n*th *Fermat number*, F_n, by $F_n = 2^{2^n} + 1$ for $n = 0, 1, 2, 3, \ldots$. The first few Fermat numbers are $F_0 = 3$, $F_1 = 5$, $F_2 = 17$, $F_3 = 257$.

    (a) Prove by induction that $F_0 \cdot F_1 \cdots F_{n-1} + 2 = F_n$, for $n \geq 1$.
    (b) Use the formula in part (a) to prove that there are an infinite number of primes, by showing that no two Fermat numbers have any prime factors in common.
    [Hint: For each F_n, let p_n be a prime divisor of F_n and show that $p_{n_1} \neq p_{n_2}$ if $n_1 \neq n_2$.]

15. The sequence of Fibonacci numbers is defined as follows: $x_1 = 1$, $x_2 = 1$, and, for $n > 2$, $x_n = x_{n-1} + x_{n-2}$. Prove that

$$x_n = \frac{1}{\sqrt{5}} \left[\left(\frac{1 + \sqrt{5}}{2} \right)^n - \left(\frac{1 - \sqrt{5}}{2} \right)^n \right]$$

for every natural number n.

[Hint: Use the fact that $x = \frac{1+\sqrt{5}}{2}$ and $x = \frac{1-\sqrt{5}}{2}$ both satisfy $1 + x = x^2$.]
16. Prove the following generalization of Theorem 2.1.6:

Theorem. *For each natural number n, consider a square consisting of 2^{2n} smaller squares (i.e., a $2^n \times 2^n$ square). If any one of the smaller squares is removed from the large square (not necessarily from the corner), then the resulting region can be completely covered by trominos (each made up of three small squares of the same size as the small squares in the region) in such a way that the trominos do not overlap.*

Chapter 3
Modular Arithmetic

Consider the number obtained by adding 3 to the number consisting of 2 to the power 3,000,005; that is, consider the number $3 + 2^{3,000,005}$. This is a very big number. No computer that presently exists, or is even conceivable, would have sufficient capacity to display all the digits in that number.

When that huge number is divided by 7, what remainder is left? You can't use your calculator, or any computer, because they can't count that high. However, this and similar questions are easily answered using a kind of "calculus" of divisibility and remainders that is called *modular arithmetic*. Another application of this study will be to prove that a natural number is divisible by 9 if and only if the sum of its digits is divisible by 9. The mathematics that we develop in this chapter has numerous other applications, including, for example, providing the basis for an extremely powerful method for sending coded messages (see Chapter 6).

3.1 The Basics

Recall that we say that the integer n is *divisible* by the integer m if there exists an integer q such that $n = mq$. In this situation, we also say that m is a *divisor* of n, or m is a *factor* of n.

The fundamental definition for modular arithmetic is the following.

Definition 3.1.1. For any fixed natural number m greater than 1, we say that the integer a *is congruent to the integer b modulo m* if $a - b$ is divisible by m. We use the notation $a \equiv b \pmod{m}$ to denote this relationship. The number m in this notation is called the *modulus*.

Here are a few examples:

$14 \equiv 8 \pmod 3$, since $14 - 8 = 6$ is divisible by 3.
$252 \equiv 127 \pmod 5$, since $252 - 127 = 125$ is divisible by 5.
$3 \equiv -11 \pmod 7$, since $3 - (-11) = 14$ is divisible by 7.

D. Rosenthal et al., *A Readable Introduction to Real Mathematics*,
Undergraduate Texts in Mathematics, DOI 10.1007/978-3-319-05654-8_3,
© Springer International Publishing Switzerland 2014

Congruence shares an important property with equality.

Theorem 3.1.2. *If $a \equiv b$ (mod m) and $b \equiv c$ (mod m), then $a \equiv c$ (mod m).*

Proof. The hypothesis states that $a - b$ and $b - c$ are both divisible by m; that is, there are integers t and s such that $a - b = tm$ and $b - c = sm$. Thus, $a - c = a - b + b - c = tm + sm = (t + s)m$. In other words, $a - c$ is divisible by m. By definition, then, $a \equiv c$ (mod m).   □

The theorem just proven shows that we can replace numbers in a congruence modulo m by any numbers congruent to them modulo m.

Although the modulus m must be bigger than 1, there is no such restriction on the integers a and b; they could even be negative. In the case where a and b are positive integers, the relationship $a \equiv b$ (mod m) can be expressed in more familiar terms.

Theorem 3.1.3. *When a and b are nonnegative integers, the relationship $a \equiv b$ (mod m) is equivalent to a and b leaving equal remainders upon division by m.*

Proof. Consider dividing m into a; if it "goes in evenly," then m is a divisor of a and the remainder r is 0. In any case, there are nonnegative integers q and r such that $a = qm + r$; q is the quotient and r is the remainder. The nonnegative number r is less than m, since it is the remainder. Similarly, divide b by m, getting $b = q_0 m + r_0$. This yields

$$a - b = (qm + r) - (q_0 m + r_0) = m(q - q_0) + (r - r_0)$$

If $r = r_0$, then $a - b$ is obviously divisible by m, so $a \equiv b$ (mod m). Conversely, if r is not equal to r_0, note that $r - r_0$ cannot be a multiple of m. (This follows from the fact that r and r_0 are both nonnegative numbers which are strictly less than m.) Thus, $a - b$ is a multiple of m plus a number that is not a multiple of m, and therefore $a - b$ is not a multiple of m. That is, it is not the case that $a \equiv b$ (mod m).   □

A special case of the above theorem is that a positive number is congruent modulo m to the remainder it leaves upon division by m. The possible remainders upon division by a given natural number m are $0, 1, 2, \ldots, m - 1$.

Theorem 3.1.4. *For a given modulus m, each integer is congruent to exactly one of the numbers in the set $\{0, 1, 2, \ldots, m - 1\}$.*

Proof. Let a be an integer. If a is positive, the result follows from the fact, discussed above, that a is congruent to the remainder it leaves upon division by m. If a is not positive, choosing t big enough would make $tm + a$ positive. For such a t, $tm + a$ is congruent to the remainder it leaves upon division by m. But also $tm + a \equiv a$ (mod m). It follows from Theorem 3.1.2 that a is congruent to the remainder that $tm + a$ leaves upon division by m. An integer cannot be congruent to two different numbers in the given set $\{0, 1, 2, \ldots, m - 1\}$, since no two numbers in the set are congruent to each other.   □

For a fixed modulus, congruences have some properties that are similar to those for equations.

Theorem 3.1.5. *If $a \equiv b$ (mod m) and $c \equiv d$ (mod m), then*

(i) $(a + c) \equiv (b + d)$ (mod m), and
(ii) $ac \equiv bd$ (mod m).

Proof. To prove (i), note that $a \equiv b$ (mod m) means that $a - b = sm$ for some integer s. Similarly, $c - d = tm$ for some integer t. The conclusion we are trying to establish is equivalent to the assertion that $(a + c) - (b + d)$ is a multiple of m. But $(a + c) - (b + d) = (a - b) + (c - d)$, which is equal to $sm + tm = (s + t)m$, so the result follows.

To prove (ii), note that from $a - b = sm$ and $c - d = tm$, we get $a = b + sm$ and $c = d + tm$, so

$$ac = (b + sm)(d + tm) = bd + btm + smd + stm^2$$

It follows that $ac - bd = m(bt + sd + stm)$, so $ac - bd$ is a multiple of m and the result is established.   □

Theorem 3.1.5 tells us that congruences are similar to equations in that you can add congruent numbers to both sides of a congruence or multiply both sides of a congruence by congruent numbers and preserve the congruence, as long as all the congruences are with respect to the same fixed modulus.

For example, since $3 \equiv 28$ (mod 5) and $17 \equiv 2$ (mod 5), it follows that $20 \equiv 30$ (mod 5) and $51 \equiv 56$ (mod 5).

Here is another example: $8 \equiv 1$ (mod 7), so $8^2 \equiv 1^2$ (mod 7), or $8^2 \equiv 1$ (mod 7). It follows that $8^2 \cdot 8 \equiv 1 \cdot 1$ (mod 7), or $8^3 \equiv 1$ (mod 7). In fact, all positive integer powers of 8 are congruent to 1 modulo 7. This is a special case of the next result.

Theorem 3.1.6. *If $a \equiv b$ (mod m), then, for every natural number n, $a^n \equiv b^n$ (mod m).*

Proof. We use mathematical induction. The case $n = 1$ is the hypothesis. Assume that the result is true for $n \geq 1$; that is, $a^n \equiv b^n$ (mod m). Since $a \equiv b$ (mod m), using part (ii) of Theorem 3.1.5 gives $a \cdot a^n \equiv b \cdot b^n$ (mod m), or $a^{n+1} \equiv b^{n+1}$ (mod m).   □

3.2   Some Applications

We can use the above to easily solve the problem that we mentioned at the beginning of this chapter: what is the remainder left when $3 + 2^{3,000,005}$ is divided by 7?

First note that $2^3 = 8$ is congruent to 1 modulo 7. Therefore, by Theorem 3.1.6, $(2^3)^{1,000,000}$ is congruent to $1^{1,000,000}$, which is 1 modulo 7. Thus $2^{3,000,000} \equiv 1$ (mod 7). Since $2^5 \equiv 4$ (mod 7) and $2^{3,000,005} = 2^{3,000,000} \cdot 2^5$, it follows that

$2^{3,000,005} \equiv 4$ (mod 7). Thus, $3 + 2^{3,000,005} \equiv (3 + 4)$ (mod 7) $\equiv 0$ (mod 7). Therefore, 7 is a divisor of $3 + 2^{3,000,005}$. In other words, the remainder that is left when $3 + 2^{3,000,005}$ is divided by 7 is 0.

Let's look at the next question we mentioned at the beginning of this chapter, the relationship between divisibility by 9 of a number and divisibility by 9 of the sum of the digits of the number. To illustrate, we begin with a particular example. Consider the number 73,486. What that really means is

$$7 \cdot 10^4 + 3 \cdot 10^3 + 4 \cdot 10^2 + 8 \cdot 10 + 6$$

Note that 10 is congruent to 1 modulo 9, so 10^n is congruent to 1 modulo 9 for every natural number n. Thus, $a \cdot 10^n \equiv a$ (mod 9) for every a and every n. It follows that $7 \cdot 10^4 + 3 \cdot 10^3 + 4 \cdot 10^2 + 8 \cdot 10 + 6$ is congruent to $(7+3+4+8+6)$ modulo 9. Thus, the number 73,486 and the sum of its digits are congruent to each other modulo 9 and therefore leave the same remainders upon division by 9. The general theorem is the following.

Theorem 3.2.1. *Every natural number is congruent to the sum of its digits modulo 9. In particular, a natural number is divisible by 9 if and only if the sum of its digits is divisible by 9.*

Proof. If n is a natural number, then we can write it in terms of its digits in the form $a_k a_{k-1} a_{k-2} \ldots a_1 a_0$ (note that this is a listing of digits, not a product of digits), where each a_i is one of 0, 1, 2, 3, 4, 5, 6, 7, 8, 9 (with $a_k \neq 0$). That is, a_0 is the digit in the "1's place," a_1 is the digit in the "10's place," a_2 is the digit in the "100's place," and so on. (In the previous example, n was the number 73,486, so in that case $a_4 = 7$, $a_3 = 3$, $a_2 = 4$, $a_1 = 8$, and $a_0 = 6$.) This really means that

$$n = a_k \cdot 10^k + a_{k-1} \cdot 10^{k-1} + a_{k-2} \cdot 10^{k-2} + \cdots + a_2 \cdot 10^2 + a_1 \cdot 10 + a_0$$

As shown above, $10 \equiv 1$ (mod 9) implies $10^i \equiv 1$ (mod 9), for every positive integer i. Therefore, n is congruent to $(a_k + a_{k-1} + a_{k-2} + \cdots + a_1 + a_0)$ modulo 9. Thus, n and the sum of its digits leave the same remainders upon division by 9. In particular, n is divisible by 9 if and only if the sum of its digits is divisible by 9. □

Congruence equations with small moduli can easily be solved by just trying all possibilities.

Example 3.2.2. Find a solution to the congruence $5x \equiv 11$ (mod 19).

Solution. If there is a solution, then there is a solution within the set $\{0, 1, 2, \ldots, 18\}$ (by Theorem 3.1.4). If $x = 0$, then $5x = 0$, so 0 is not a solution. Similarly, for $x = 1$, $5x = 5$; for $x = 2$, $5x = 10$; for $x = 3$, $5x = 15$; and for $x = 4$, $5x = 20$. None of these are congruent to 11 (mod 19), so we have not yet found a solution. However, when $x = 6$, $5x = 30$, which is congruent to 11 (mod 19). Thus, $x \equiv 6$ (mod 19) is a solution of the congruence.

Example 3.2.3. Show that there is no solution to the congruence $x^2 \equiv 3 \pmod 5$.

Proof. If $x = 0$, then $x^2 = 0$; if $x = 1$, then $x^2 = 1$; if $x = 2$, then $x^2 = 4$; if $x = 3$, then $x^2 = 9$, which is congruent to 4 (mod 5); and if $x = 4$, then $x^2 = 16$ which is congruent to 1 (mod 5). If there was any solution, it would be congruent to one of $\{0, 1, \ldots, 4\}$ by Theorem 3.1.4. Thus, the congruence has no solution. □

3.3 Problems

Basic Exercises

1. Find a solution x to each of the following congruences. ("Solution" means integer solution.)

    (a) $2x \equiv 7 \pmod{11}$
    (b) $7x \equiv 4 \pmod{11}$
    (c) $x^5 \equiv 3 \pmod 4$

2. For each of the following congruences, either find a solution or prove that no solution exists.

    (a) $39x \equiv 13 \pmod 5$
    (b) $95x \equiv 13 \pmod 5$
    (c) $x^2 \equiv 3 \pmod 6$
    (d) $5x^2 \equiv 12 \pmod 8$
    (e) $4x^3 + 2x \equiv 7 \pmod 5$

Interesting Problems

3. Find the remainder when:

    (a) 3^{2463} is divided by 8.
    (b) 2^{923} is divided by 15.
    (c) 243^{101} is divided by 8.
    (d) $5^{2001} + (27)!$ is divided by 8.
    (e) $(-8)^{4124} + 6^{3101} + 7^5$ is divided by 3.
    (f) $7^{103} + 6^{5409}$ is divided by 3.
    (g) $5! \cdot 181 - 866 \cdot 332$ is divided by 6.

4. Is $2^{598} + 3$ divisible by 15?
5. Find a digit b such that the number $2794b2$ is divisible by 8.
6. Determine whether or not $17^{2492} + 25^{376} + 5^{782}$ is divisible by 3.
7. Suppose that 7^{22} is written out in the ordinary way. What is its last digit?

8. Determine whether or not the following congruence has a natural number solution:

$$5^x + 3 \equiv 5 \ (\text{mod } 100)$$

9. Prove that $n^2 - 1$ is divisible by 8, for every odd integer n.
10. Prove that a natural number is divisible by 3 if and only if the sum of its digits is divisible by 3.
11. Prove that $x^5 \equiv x \ (\text{mod } 10)$, for every integer x. (This shows that x^5 and x have the same units' digit for every integer x.)
12. Suppose a number is written in decimal notation as $abba$, where a and b are integers between 1 and 9. Prove that this number is divisible by 11.
13. Find the units' digit of 27493^{6782}.
14. Show that if m is a natural number and a is a negative integer, then there exists an r with $0 \leq r \leq m - 1$ and an integer q such that $a = qm + r$. (Cf. the proof of Theorem 3.1.3.)
15. Prove that for every pair of natural numbers m and n, m^2 is congruent to n^2 modulo $(m + n)$.

Challenging Problems

16. Prove that 5 divides $3^{2n+1} + 2^{2n+1}$, for every natural number n.
17. Prove that 7 divides $8^{2n+1} + 6^{2n+1}$, for every natural number n.
18. Prove that a natural number that is congruent to 2 modulo 3 has a prime factor that is congruent to 2 modulo 3.
19. If m is a natural number greater than 1 and is not prime, then we know that $m = ab$, where $1 < a < m$ and $1 < b < m$. Show that there is no integer x such that $ax \equiv 1 \ (\text{mod } m)$. (That is, a has no *multiplicative inverse modulo m*. The situation is different if m is prime: see Problem 7 in Chapter 4.)
20. Prove that 133 divides $11^{n+1} + 12^{2n-1}$, for every natural number n.
21. A natural number r less than or equal to $m - 1$ is called a *quadratic residue modulo m* if there is an integer x such that $x^2 \equiv r \ (\text{mod } m)$. Determine all the quadratic residues modulo 11.
22. Show that there do not exist natural numbers x and y such that $x^2 + y^2 = 4003$. [Hint: Begin by determining which of the numbers $\{0, 1, 2, 3\}$ can be congruent to $x^2 \ (\text{mod } 4)$.]
23. Discover and prove a theorem determining whether a natural number is divisible by 11, in terms of its digits.
24. Prove that there are an infinite number of primes of the form $4k + 3$ with k a natural number. [Hint: If $p_1, p_2, \ldots, p_n$ are n such primes, show that $(4 \cdot p_1 \cdot p_2 \cdots p_n) - 1$ has at least one prime divisor of the given form.]

25. Prove that there are an infinite number of primes of the form $6k + 5$ with k a natural number.

26. Prove that every prime number greater than 3 differs by 1 from a multiple of 6.

27. Show that, if x, y, and z are integers such that $x^2 + y^2 = z^2$, then at least one of $\{x, y, z\}$ is divisible by 2, at least one of $\{x, y, z\}$ is divisible by 3, and at least one of $\{x, y, z\}$ is divisible by 5.

28. Let $f(x)$ be a non-constant polynomial with integer coefficients. (That is, there exists a natural number n and integers a_i such that $f(x) = a_n x^n + a_{n-1} x^{n-1} + \cdots + a_1 x + a_0$.) Let a, k, and m be integers with $m > 1$. Suppose that $f(a) \equiv k$ (mod m). Prove that $f(a + m) \equiv k$ (mod m).

29. Show that the polynomial $p(x) = x^2 - x + 41$ takes prime values for x in the set $\{0, 1, 2, \ldots, 40\}$.

30. Show that there does not exist any non-constant polynomial $p(x)$ with integer coefficients such that $p(x)$ is a prime number for all natural numbers x.

Chapter 4
The Fundamental Theorem of Arithmetic

Is $13^{217} \cdot 37^{92} \cdot 41^{15} = 19^{111} \cdot 29^{145} \cdot 43^{12} \cdot 47^{5}$?

We have seen that every natural number greater than 1 is either a prime or a product of primes. The above equation, if it was an equation, would express a number in two different ways as a product of primes. Does the representation of a natural number as a product of primes have to be unique? The answer is obviously "no" in one sense. For example, $6 = 3 \cdot 2 = 2 \cdot 3$. Thus, the same number can be written in two different ways as a product of primes if we consider different orders as "different ways." But suppose that we don't consider the ordering; must the factorization of a natural number into a product of primes be unique except for the order? For example, could the above equation hold?

In fact, every natural number other than 1 has a factorization into a product of primes and the factorization is unique except for the order. This result is so important that it is called the *Fundamental Theorem of Arithmetic*. We will give two proofs. The second proof requires a little more development and will be given later (Theorem 7.2.4). The first proof is short but tricky.

4.1 Proof of the Fundamental Theorem of Arithmetic

In order to simplify the statement of the Fundamental Theorem of Arithmetic, we use the expression "a product of primes" to include the case of a single prime number (as we did in Theorem 2.2.4).

The Fundamental Theorem of Arithmetic 4.1.1. *Every natural number greater than 1 can be written as a product of primes, and the expression of a number as a product of primes is unique except for the order of the factors.*

Proof. We have already established that every natural number greater than 1 can be written as a product of primes (see Theorem 2.2.4). That was the easy part of the Fundamental Theorem of Arithmetic; the harder part is the uniqueness. The proof

D. Rosenthal et al., *A Readable Introduction to Real Mathematics*,
Undergraduate Texts in Mathematics, DOI 10.1007/978-3-319-05654-8_4,
© Springer International Publishing Switzerland 2014

of uniqueness that we present below is a proof by contradiction. That is, we will assume that there is a natural number with more than one representation as a product of primes and derive a contradiction from this assumption, thereby showing that this assumption is incorrect.

Suppose, then, that there is at least one natural number with at least two different representations as a product of primes. By the Well-Ordering Principle (2.1.2), there would then be a smallest natural number with that property (i.e., the smallest natural number that has at least two different such representations). Let N be that smallest such number. Write out two different factorizations of N:

$$N = p_1 p_2 \cdots p_r = q_1 q_2 \cdots q_s$$

where each of the p_i and the q_j are primes (there can be repetitions of the same prime). We first claim that no p_i could be equal to any q_j. This follows from the fact that N is the smallest number with a non-unique representation, for if $p_i = q_j$ for some i and j, that common factor could be divided from both of the two different factorizations for n, producing a smaller number that has at least two different factorizations. Thus, no p_i is equal to any q_j.

Since p_1 is different from q_1, one of p_1 and q_1 is less than the other; suppose that p_1 is less than q_1. (If q_1 is less than p_1, the same proof could be repeated by simply interchanging the p's and q's.) Define M by

$$M = N - (p_1 q_2 \cdots q_s)$$

Then M is a natural number that is less than N. Substituting the product $p_1 p_2 \cdots p_r$ for N gives

$$M = (p_1 p_2 \cdots p_r) - (p_1 q_2 \cdots q_s) = p_1\big((p_2 \cdots p_r) - (q_2 \cdots q_s)\big)$$

from which it follows that p_1 divides M. In particular, M is not 1. Since M is less than N, M has a unique factorization into primes.

Substituting the product $q_1 q_2 \cdots q_s$ for N in the definition of M gives a different expression:

$$M = (q_1 q_2 \cdots q_s) - (p_1 q_2 \cdots q_s) = (q_1 - p_1)(q_2 \cdots q_s)$$

The unique factorization of M into primes can thus be obtained by writing the unique factorization of $q_1 - p_1$ followed by the product $q_2 \cdots q_s$. On the other hand, the fact that p_1 is a divisor of M implies that p_1 must appear in the factorization of M into primes. Since p_1 is distinct from each of $\{q_2, \ldots, q_s\}$, it follows that p_1 must occur in the factorization of $q_1 - p_1$ into primes. Thus, $q_1 - p_1 = p_1 k$, for some natural number k. It follows that $q_1 = p_1 + p_1 k = p_1(1 + k)$, which shows that q_1 is divisible by p_1. Since p_1 and q_1 are distinct primes, this is impossible. Hence, the assumption that there is a natural number with two distinct factorizations leads to a contradiction, so factorizations into primes are unique. $\square$

The Fundamental Theorem of Arithmetic gives a so-called "canonical form" for expressing each natural number greater than 1.

Corollary 4.1.2. *Every natural number n greater than 1 has a canonical factorization into primes; that is, n has a unique representation of the form $n = p_1^{\alpha_1} p_2^{\alpha_2} \cdots p_n^{\alpha_n}$, where each p_i is a prime, p_i is less than p_{i+1} for each i, and each α_i is a natural number.*

Proof. To see this, simply factor the given number as a product of primes and then collect all occurrences of the smallest prime together, then all the occurrences of the next smallest prime, and so on.   □

For example, the canonical form of $60,368$ is $2^4 \cdot 7^3 \cdot 11$. The canonical form of 19 is simply 19.

As we will see, the following corollary of the Fundamental Theorem of Arithmetic is very useful. (If the corollary below is independently established, then it is easy to derive the Fundamental Theorem of Arithmetic from it. In fact, most presentations of the proof of the Fundamental Theorem of Arithmetic use this approach rather than the shorter but trickier proof that we gave above. We will present such a proof later (Theorem 7.2.4).)

Corollary 4.1.3. *If p is a prime number and a and b are natural numbers such that p divides ab, then p divides at least one of a and b. (That is, if a prime divides a product, then it divides at least one of the factors.)*

Proof. Since p divides ab, there is some natural number d such that $ab = pd$. The unique factorization of ab into primes therefore contains the prime p and all the primes that divide d. On the other hand, a and b each have unique factorizations into primes. Let the canonical factorization of a be $q_1^{\alpha_1} q_2^{\alpha_2} \cdots q_m^{\alpha_m}$ and of b be $r_1^{\beta_1} r_2^{\beta_2} \cdots r_n^{\beta_n}$. Then,

$$ab = (q_1^{\alpha_1} q_2^{\alpha_2} \cdots q_m^{\alpha_m})(r_1^{\beta_1} r_2^{\beta_2} \cdots r_n^{\beta_n})$$

Since the factorization of ab into primes is unique, p must occur either as one of the q_i's, in which case p divides a, or as one of the r_j's, in which case p divides b. Thus, p divides at least one of a and b, and the corollary is established.   □

It should be noted that this corollary is not generally true for divisors that are not prime. For example, 18 divides $3 \cdot 12$, but 18 does not divide 3 and 18 does not divide 12.

4.2 Problems

Basic Exercises

1. Find the canonical factorization into primes of each of the following:

(a) 52   (e) $122 \cdot 54$
(b) 72   (f) 112
(c) 47   (g) 224
(d) 625   (h) $112 + 224$

2. Find natural numbers x, y, and z such that

   (a) $3^x \cdot 100 \cdot 5^y = 9 \cdot 10^z \cdot 5$
   (b) $50 \cdot 2^y \cdot 7^z = 5^x \cdot 2^3 \cdot 14$

3. Show that if p is a prime number and $a_1, a_2, \ldots, a_n$ are natural numbers such that p divides the product $a_1 a_2 \cdots a_n$, then p divides a_i for at least one a_i.

4. Show that if p is a prime number and a and n are natural numbers such that p divides a^n, then p divides a.

Interesting Problems

5. Find the smallest natural numbers x and y such that

   (a) $7^2 x = 5^3 y$
   (b) $2^5 x = 10^2 y$
   (c) $127 x = 54 y$

6. Find nonnegative integers w, x, y, and z such that

$$17^2 25^2 2^z = 10^x 34^y 7^w$$

Challenging Problems

7. Suppose that p is a prime number and p does not divide a. Prove that the congruence $ax \equiv 1 \pmod{p}$ has a solution. (This proves that a has a *multiplicative inverse modulo p*.)

8. Prove that a natural number m greater than 1 is prime if m has the property that it divides at least one of a and b whenever it divides ab.

9. Prove that $x^2 \equiv 1 \pmod{p}$ implies $x \equiv 1 \pmod{p}$ or $x \equiv (p-1) \pmod{p}$, for every prime p.

10. Suppose that a and b are natural numbers whose prime factorizations have no primes in common (the pair a, b is then said to be *relatively prime*; see Definition 7.2.1). Show that for any natural number m, the product ab divides m if each of a and b divides m.

11. Using the result of Problem 10:

    (a) Prove that 42 divides $n^7 - n$, for every natural number n.
    (b) Prove that 21 divides $3n^7 + 7n^3 + 11n$, for every natural number n.

Chapter 5
Fermat's Theorem and Wilson's Theorem

We've seen that we can add or multiply "both sides" of a congruence by congruent numbers and the result will be a congruence (Theorem 3.1.5). What about dividing both sides of a congruence by the same natural number? For the result to have a chance of being a congruence, the divisor must divide evenly into both sides of the congruence so that the result involves only integers, not fractions (congruences are only defined for integers). On the other hand, even that condition is not sufficient to ensure that the result will be a congruence. For example, $6 \cdot 2$ is congruent to $6 \cdot 1$ modulo 3, but 2 is not congruent to 1 modulo 3. This is not a surprising example, since 6 is congruent to 0 modulo 3, so "dividing both sides" of the above congruence by 6 is like dividing by 0, which gives wrong results for equations as well. However, there are also examples where dividing both sides of a congruence by a number that is not congruent to 0 leads to results that are not congruent. For example, $12 \cdot 3$ is congruent to $24 \cdot 3$ modulo 9, but 12 is not congruent to 24 modulo 9, in spite of the fact that 3 is not congruent to 0 modulo 9.

5.1 Fermat's Theorem

There are important cases in which we can divide both sides of a congruence and be assured that the result is a congruence.

Theorem 5.1.1. *If p is a prime and a is not divisible by p, and if $ab \equiv ac$ (mod p), then $b \equiv c$ (mod p). (That is, we can divide both sides of a congruence modulo a prime by any natural number that divides both sides of the congruence and is not divisible by the prime.)*

Proof. We are given that p divides $ab - ac$. This is the same as saying that p divides $a(b - c)$. Corollary 4.1.3 shows that since p divides $a(b - c)$, p must also divide either a or $b - c$. Since the hypothesis states that a is not divisible by p, this implies that $b - c$ must be divisible by p. That is the same as saying $b \equiv c$ (mod p). $\quad\square$

D. Rosenthal et al., *A Readable Introduction to Real Mathematics*,
Undergraduate Texts in Mathematics, DOI 10.1007/978-3-319-05654-8__5,
© Springer International Publishing Switzerland 2014

Consider any given prime number p. The possible remainders when a natural number is divided by p are the numbers $\{0, 1, \ldots, p - 1\}$. By Theorem 3.1.4, no two of these numbers are congruent to each other and every natural number (in fact, every integer) is congruent modulo p to one of those numbers. An integer is divisible by p if and only if it is congruent to 0 modulo p. Thus, each integer that is not divisible by p is congruent to exactly one of the numbers in the set $\{1, 2, \ldots, p - 1\}$. This is the basis for the proof of the following beautiful, and very useful, theorem.

Fermat's Theorem 5.1.2. *If p is a prime number and a is any natural number that is not divisible by p, then $a^{p-1} \equiv 1 \pmod{p}$.*

Proof. Let p be any prime number and let a be any natural number that is not divisible by p. Consider the set of numbers $\{a \cdot 1, a \cdot 2, \ldots, a \cdot (p - 1)\}$. First note that no two of those numbers are congruent to each other, for if $am \equiv an$ $\pmod{p}$, then, by Theorem 5.1.1, $m \equiv n \pmod{p}$. Since no two of the numbers in the set $\{1, 2, \ldots, p - 1\}$ are congruent to each other, this shows that the same is true of numbers in the set $\{a \cdot 1, a \cdot 2, \ldots, a \cdot (p - 1)\}$. Also note that each of the numbers in the set $\{a \cdot 1, a \cdot 2, \ldots, a \cdot (p - 1)\}$ is congruent to one of the numbers in $\{1, 2, \ldots, p - 1\}$ since no number in either set is divisible by p. Thus, the numbers in the set $\{a \cdot 1, a \cdot 2, \ldots, a \cdot (p - 1)\}$ are congruent, in some order, to the numbers in the set $\{1, 2, \ldots, p - 1\}$. This implies that the product of all of the numbers in the set $\{a \cdot 1, a \cdot 2, \ldots, a \cdot (p - 1)\}$ is congruent modulo p to the product of all the numbers in $\{1, 2, \ldots, p - 1\}$. Thus, $a \cdot 1 \cdot a \cdot 2 \cdots a \cdot (p - 1)$ is congruent to $1 \cdot 2 \cdot 3 \cdots (p - 1)$ modulo p. Since the number a occurs $p - 1$ times in this congruence, this yields $a^{p-1}(1 \cdot 2 \cdot 3 \cdots (p - 1)) \equiv (1 \cdot 2 \cdot 3 \cdots (p - 1))$ $\pmod{p}$. Clearly, p does not divide $1 \cdot 2 \cdot 3 \cdots (p - 1)$ (by repeated application of Corollary 4.1.3). Thus, by Theorem 5.1.1, we can "divide" both sides of the above congruence by $1 \cdot 2 \cdot 3 \cdots (p - 1)$, yielding $a^{p-1} \equiv 1 \pmod{p}$.   □

As we shall see, Fermat's Theorem has important applications, including in establishing a method for sending coded messages. It is also sometimes useful to apply Fermat's Theorem to specific cases. For example, $88^{100} - 1$ is divisible by 101. (Don't try to verify this on your calculator!)

The following corollary of Fermat's Theorem is sometimes useful since it doesn't require that a not be divisible by p.

Corollary 5.1.3. *If p is a prime number and a is any natural number, then $a^p \equiv a$ $\pmod{p}$.*

Proof. If p does not divide a, then Fermat's Theorem states that $a^{p-1} \equiv 1 \pmod{p}$. Multiplying both sides of this congruence by a gives the result in this case. On the other hand, if p does divide a, then p also divides a^p, so a^p and a are both congruent to 0 mod p.   □

Definition 5.1.4. A *multiplicative inverse modulo p* for a natural number a is a natural number b such that $ab \equiv 1 \pmod{p}$.

Fermat's Theorem is one way of showing that all natural numbers that are not multiples of a given prime p have multiplicative inverses modulo p.

Corollary 5.1.5. *If p is a prime and a is a natural number that is not divisible by p, then there exists a natural number x such that $ax \equiv 1 \pmod{p}$.*

Proof. In the case where p is the prime 2, each such a must be congruent to 1 modulo 2, so we can take $x = 1$. If p is greater than 2, then, for each given a, let $x = a^{p-2}$. Then $ax = a \cdot a^{p-2} = a^{p-1}$ and, by Fermat's Theorem, $a^{p-1} \equiv 1 \pmod{p}$. □

The following lemma is needed in the proof of Wilson's Theorem (5.2.1).

Lemma 5.1.6. *If a and c have the same multiplicative inverse modulo p, then a is congruent to c modulo p.*

Proof. Suppose $ab \equiv 1 \pmod{p}$ and $cb \equiv 1 \pmod{p}$. Then multiplying the second congruence on the right by a yields $cba \equiv a \pmod{p}$ and, since $ba \equiv 1 \pmod{p}$, this gives $c \equiv a \pmod{p}$. □

It turns out to be interesting and useful to know which natural numbers are congruent to their own inverses modulo p. If x is such a number, then $x \cdot x \equiv 1 \pmod{p}$. In other words, such an x is a solution to the congruence $x^2 \equiv 1 \pmod{p}$, or $x^2 - 1 \equiv 0 \pmod{p}$. The solutions of the equations $x^2 - 1 = 0$ are $x = 1$ and $x = -1$. The solutions of the congruence are similar.

Theorem 5.1.7. *If p is a prime number and x is an integer satisfying $x^2 \equiv 1 \pmod{p}$, then either $x \equiv 1 \pmod{p}$ or $x \equiv p-1 \pmod{p}$. (Note that $p-1 \equiv -1 \pmod{p}$).*

Proof. If $x^2 \equiv 1 \pmod{p}$, then, by definition, p divides $x^2 - 1$. But $x^2 - 1 = (x - 1)(x + 1)$. Since p divides $x^2 - 1$, Corollary 4.1.3 implies that p divides at least one of $x - 1$ and $x + 1$. If p divides $x - 1$, then $x \equiv 1 \pmod{p}$. If p divides $x + 1$, then $x \equiv -1 \pmod{p}$, or $x \equiv p - 1 \pmod{p}$. □

5.2 Wilson's Theorem

As we now show, these considerations lead to a proof of Wilson's Theorem, a theorem that is very beautiful although it is considerably less famous and much less useful than Fermat's Theorem.

Wilson's Theorem 5.2.1. *If p is a prime number, then $(p - 1)! + 1 \equiv 0 \pmod{p}$. (In other words, if p is prime, then p divides $(p - 1)! + 1$.)*

Proof. First note that the theorem is obviously true when $p=2$; in this case, it states $(1 + 1) \equiv 0 \pmod{2}$. As we indicated above, a multiplicative inverse of an integer x modulo p is an integer y such that $xy \equiv 1 \pmod{p}$. As we have seen, every number in the set $\{1, 2, \ldots, p - 1\}$ is distinct modulo p, and, by Corollary 5.1.5, each has a multiplicative inverse modulo p. Since no multiplicative inverse can be divisible by p, the multiplicative inverse of each number in $\{1, 2, \ldots, p - 1\}$ is congruent to one of the numbers in $\{1, 2, \ldots, p - 1\}$. By the previous theorem (5.1.7), the only numbers in the set $\{1, 2, \ldots, p - 1\}$ that are congruent to their own multiplicative inverses are the numbers 1 and $p - 1$. Leave those two numbers aside for the moment. Note that if y is a multiplicative inverse of x, then x is a multiplicative inverse of y. Thus, the numbers in the set $\{2, 3, \ldots, p - 2\}$ each have multiplicative inverses in that same set, and each number in that set differs from its multiplicative inverse. By Lemma 5.1.6, no two numbers in the set can have the same inverse. Therefore, we can arrange the numbers in the set $\{2, 3, \ldots, p - 2\}$ in pairs consisting of a number and its multiplicative inverse. Since the product of a number and its multiplicative inverse is congruent to 1 modulo p, the product of all the numbers in the set $\{2, 3, \ldots, p - 2\}$ is congruent to 1 modulo p. Thus, $2 \cdot 3 \cdots (p-2) \equiv 1 \pmod{p}$. Multiplying both sides by 1 gives $1 \cdot 2 \cdot 3 \cdots (p-2) \equiv 1 \pmod{p}$. Now $p-1 \equiv -1 \pmod{p}$, so $1 \cdot 2 \cdots (p-2) \cdot (p-1) \equiv 1 \cdot (-1) \pmod{p}$. In other words, $(p - 1)! \equiv -1 \pmod{p}$, which yields $(p - 1)! + 1 \equiv 0 \pmod{p}$. $\qquad\square$

Theorem 5.2.2. *If m is a composite number larger than 4, then $(m - 1)! \equiv 0 \pmod{m}$ (so that $(m - 1)! + 1 \equiv 1 \pmod{m}$).*

Proof. Let m be any composite number larger than 4. We must show that $(m - 1)!$ is divisible by m. If $m = ab$, with a different from b and both less than m, then a and b each occur as distinct factors in $(m - 1)!$. Thus, $m = ab$ is a factor of $(m - 1)!$, so $(m - 1)!$ is congruent to 0 modulo m. The only composite numbers less than m that cannot be written as a product of two distinct natural numbers less than m are those numbers that are squares of primes. (To see this, use the fact that every composite can be written as a product of primes.) Thus, the only remaining case to prove is when $m = p^2$ for some prime p. In this case, if m is larger than 4, then p is a prime bigger than 2. In that case, p^2 is greater than $2p$. Thus, $p^2 - 1$ is greater than or equal to $2p$, so $(p^2 - 1)!$ contains the factor $2p$ as well as the factor p. Thus, $(p^2 - 1)!$ contains the product $2p^2$. In particular, $(m - 1)!$ is divisible by $m = p^2$. $\qquad\square$

The following combines Wilson's Theorem and its converse.

Theorem 5.2.3. *If m is a natural number other than 1, then $(m - 1)! + 1 \equiv 0 \pmod{m}$ if and only if m is a prime number.*

Proof. This follows immediately from Wilson's Theorem when m is prime and from the previous theorem for composite m in all cases except for $m = 4$. If $m = 4$, then $(m - 1)! + 1 = 3! + 1 = 7$, which is not congruent to 0 modulo 4, so the theorem holds for all m. $\qquad\square$

It might be thought that Wilson's Theorem would provide a good way to check whether or not a given number m is prime: simply see whether m divides $(m - 1)! + 1$. However, the fact that $(m - 1)!$ is so much larger than m makes this a very impractical way of testing primality for large values of m.

5.3 Problems

Basic Exercises

1. Find the remainder when 24^{103} is divided by 103.
2. Find a solution x to each of the following congruences:

   (a) $2^x \equiv 1 \pmod{103}$
   (b) $16! \cdot x \equiv 5 \pmod{17}$

3. Find the remainder when $99^{100} - 1$ is divided by 101.

Interesting Problems

4. Suppose that p is a prime greater than 2 and $a \equiv b^2 \pmod{p}$ for some natural number b that is not divisible by p. Prove that $a^{\frac{p-1}{2}} \equiv 1 \pmod{p}$.
5. Find three different prime factors of $10^{12} - 1$.
6. Let p be a prime number. Prove that $1^2 \cdot 2^2 \cdot 3^2 \cdots (p-1)^2 - 1$ is divisible by p.
7. For each of the following congruences, either find a solution or prove that no solution exists.

   (a) $102! \cdot x + x \equiv 4 \pmod{103}$
   (b) $x^{16} - 2 \equiv 0 \pmod{17}$

8. Find the remainder when:

   (a) $(9! \cdot 16 + 4311)^{8603}$ is divided by 11.
   (b) $42! + 7^{28} + 66$ is divided by 29.

9. If a is a natural number and p is a prime number, show that $a^p + a \cdot (p-1)!$ is divisible by p.
10. Find the remainder that $100 + 2^{33} + 16! + 29!$ leaves upon division by 19.

Challenging Problems

11. Show that a natural number $n > 1$ is prime if and only if n divides $(n-2)! - 1$.
12. Show that if p is a prime number and a and b are natural numbers, then

$$(a + b)^p \equiv a^p + b^p \pmod{p}$$

13. For which prime numbers p is $(p-2)! \equiv 1 \pmod{p}$?
14. Prove that for all primes $p > 3$, $2 \cdot (p-3)! \equiv -1 \pmod{p}$.

15. Is there a prime number p such that $(p-1)! + 6$ is divisible by p?
16. Find all prime numbers p such that p divides $(p-2)! + 6$.
17. Suppose $2^k + 1$ is a prime number. Prove that k has no prime divisors other than 2.

    [Hint: If $k = ab$ with b odd, consider $2^k + 1$ modulo $2^a + 1$.]
18. Prove that $a^{q-1} \equiv 1 \pmod{pq}$ if p and q are distinct primes such that $p-1$ divides $q-1$ and neither p nor q divides a.

Chapter 6
Sending and Receiving Secret Messages

As early as ancient times people have devised ways of sending secret messages to each other. Much of the original interest was for military purposes: commanders of one section of an army wanted to send messages to commanders of other sections of their army in such a way that the message could not be understood by enemy soldiers who might intercept it.

Some of the current interest in secret messages is still for military and similarly horrible purposes. However, there are also many other kinds of situations in which it is important to be able to send secret messages. For example, a huge amount of information is communicated via the internet. It is important that some of that information remain private, known only to the sender and recipient. One common situation is making withdrawals from bank accounts over the internet. If someone else was able to intercept the information being sent, that interceptor could transfer funds from the sender's bank account to the interceptor's bank account. There are many other commercial and personal communications that are sent electronically that people wish to keep secret.

"Cryptography" refers to techniques for reconfiguring messages so that they cannot be understood except by the intended recipient. *Encrypting* a message is the process of reconfiguring it; *decrypting* is the process of obtaining the original message from the encrypted one. For a method of cryptography to be useful, it must be the case that it would be virtually impossible (or at least extremely difficult) for anyone other than the intended recipient to be able to decrypt the messages.

A fundamental problem is that the intended recipient must have the information that is needed to decrypt encrypted messages. If the sender has to send the decrypting information to the recipient, unintended interceptors (e.g., someone who wants to transfer your money to his or her bank account) might get access to the method of decrypting, since that method is being transmitted to the intended recipient.

In most procedures for sending encrypted messages, anyone who understands the procedure for encrypting messages would also understand how to decrypt them. When such a method is employed, it is crucial that only the intended recipient of

D. Rosenthal et al., *A Readable Introduction to Real Mathematics*,
Undergraduate Texts in Mathematics, DOI 10.1007/978-3-319-05654-8_6,
© Springer International Publishing Switzerland 2014

messages obtains the knowledge of how to decrypt the messages. It can be very difficult to get the method of decrypting to the intended recipient.

The techniques of encrypting and decrypting messages for a given procedure are called the "keys" for that procedure. There must be a "key" for encrypting messages and a "key" for decrypting them.

Beginning in the 1940s, many people wondered whether there could be *public key cryptography*. That means, a method of doing cryptography that has the property that everyone in the world (the "public") is told how to send the recipient an encrypted message. On the other hand, the recipient must be the only one who can decrypt messages sent using that procedure. That is, *public key cryptography* refers to methods of sending messages that allow the person who wishes to receive messages to publicly announce the way messages should be encrypted in such a way that only the person making the announcement can decrypt a message. This seems to be impossible. If people know how to encrypt messages, won't they necessarily also be able to figure out how to decrypt them, just by reversing the encrypting procedure?

6.1   The RSA Method

It was only in the 1970s that a method for public key cryptography was discovered. To actually use this method requires employing very large numbers. Thus, this method would not be feasible without computers. On the other hand, the only mathematics that is required to establish that the method works is Fermat's Theorem (5.1.2). This method is called "RSA" after three of the people who played important roles in discovering it, Ron Rivest, Adi Shamir, and Leonard Adleman.

Here is an outline of the method. The recipient announces to the entire world the following way to send messages. If you want to send a message, the first thing that you must do is to convert the message into a natural number. There are many ways of doing that; here is a rough description of one possibility. Write your message out as sentences in, say, the English language. Then convert the sentences into a natural number as follows. Let $a = 11$, $b = 12$, $c = 13$, ..., $z = 36$. Let 37 represent a space. Let 38 represent a period, 39 a comma, 40 a semicolon, 41 a full colon, 42 an exclamation point, and 43 an apostrophe. If desired, other symbols could be represented by other two-digit natural numbers. Convert your English language message into a number by replacing each of the elements of your sentences by their corresponding numbers in the order that they appear. For any substantial message, this will result in a large natural number. Everyone would be able to reconstruct the English language message from that number if this procedure was known to them. For example, the sentence

Public key cryptography is neat.

would be represented by the number

26311222191337211535371328352630251728112618353719293724151l3038

Furthermore, if you read the rest of this chapter

3525314322223721242533373318353719303733252821 2938

The RSA technique is a method for encrypting and decrypting numbers. Both the recipient and those who send messages must use computers to do the computations that are required; the numbers involved in any application of the technique that could realistically protect messages are much too large for the computations to be done by hand.

RSA encryption proceeds as follows. The person who wishes to receive messages, the recipient, chooses two very large prime numbers p and q that are different from each other, and then defines N to be pq. The recipient publicly announces the number N. However, the recipient keeps p and q secret. If p and q are large enough, there is no way that anyone other than the recipient could find p or q simply from knowing N; factoring very large numbers is beyond the capacity of even the most powerful computers. There are some very large known prime numbers; such can easily be chosen so that the resulting $N = pq$ is impossible to factor in any reasonable amount of time. The recipient announces another natural number E, which we will call the *encryptor*, in addition to N. Below we will explain ways of choosing suitable E's.

The recipient then instructs all those who wish to send messages to do the following. Write your message as a natural number as described above. Let's say that M is the number representing your message. For this method to work, M must be less than N. If M is greater than or equal to N, you could divide your message into several smaller messages, each of which correspond to natural numbers less than N. The method we shall describe only works when M is less than N.

"To send me messages," the recipient announces to the world, "take your message M and compute the remainder that M^E (i.e., M raised to the power E) leaves upon division by N, and send me that remainder."

In other words, to send a message M, the sender computes the R between 0 and N such that $M^E \equiv R \pmod{N}$. The sender then sends R to the recipient.

How can the message be decrypted? That is, how can the recipient recover the original message M from R? This will require finding a *decryptor*, which will be possible for anyone who knows the factorization of N as the product pq, but virtually impossible for anyone else. We shall see that, if E is chosen properly, there is a decryptor D such that for every integer L between 0 and N, $L^{ED} \equiv L$ $\pmod{N}$. For such a D, since $R \equiv M^E \pmod{N}$, it follows that $R^D \equiv M^{ED}$ $\pmod{N}$, and therefore, since $M^{ED} \equiv M \pmod{N}$, $R^D \equiv M \pmod{N}$. Thus, the recipient decrypts the message by finding the remainder that R^D leaves upon division by N.

Before explaining further how to find encryptors E and decryptors D and why this method works, let's look at a simple example. In this example the numbers are so small that anyone could figure out what p and q are, so this example could not realistically be used to encrypt messages. However, it illustrates the method.

Example 6.1.1. Let $p = 7$ and $q = 11$ be the primes; then $N = pq = 77$. Suppose that $E = 13$; as we shall see, there are always many possible values for E. Below we will discuss the properties that E must have. There is a technique for finding D, based on knowing p and q, that we shall describe later; that technique will produce $D = 37$ in this particular example.

In this example, the recipient announces $N = 77$ and $E = 13$ to the general public; the recipient keeps the values of p, q, and D secret.

The recipient invites the world to send messages. Suppose you want to send the message $M = 71$. Following the encryption rule, you must compute the remainder that $M^E = 71^{13}$ leaves upon division by 77. Let's compute that as follows, using some of the facts about modular arithmetic that we learned in the previous chapters. First, $71 \equiv -6 \pmod{77}$, so $M^E \equiv (-6)^{13} \pmod{77}$. Now $6^3 = 216$ and $216 \equiv -15 \pmod{77}$, so $6^6 = (6^3)^2 \equiv (-15)^2 \equiv 225 \pmod{77}$, which is congruent to $-6 \pmod{77}$. Therefore, $6^{12} \equiv (-6)^2 \equiv 36 \pmod{77}$, so $6^{13} \equiv 6 \cdot 6^{12} \equiv 6 \cdot 36 \equiv 216 \equiv -15 \pmod{77}$. Therefore, $(-6)^{13} \equiv -(6)^{13} \equiv 15 \pmod{77}$. Thus, $M^E = 71^{13} \equiv 15 \pmod{77}$. It follows that the remainder upon dividing 71^{13} by 77 is 15.

Thus, the encrypted version of your message is 15. Anyone who sees that the encrypted version is 15 would be able to discover your original message if they knew the decryptor. But the recipient is the only one who knows the decryptor.

In this special, easy, example, the recipient receives 15 and proceeds to decrypt it, using the decryptor 37, as follows. Your original message will be the remainder that 15^{37} leaves upon division by 77. Compute: $15^2 \equiv -6 \pmod{77}$. Therefore, $15^{26} \equiv (-6)^{13} \pmod{77}$, which (as we saw above) is congruent to $15 \pmod{77}$. Also, from $15^2 \equiv -6 \pmod{77}$, we obtain $15^4 \equiv 36 \pmod{77}$. Thus, $15^8 \equiv 36 \cdot 6 \cdot 6 \equiv 216 \cdot 6 \equiv (-15) \cdot 6 \equiv -90 \equiv 64 \pmod{77}$. Now $15^{37} \equiv 15^{26} \cdot 15^8 \cdot 15^3 \pmod{77}$, which is congruent to $15 \cdot 64 \cdot 15^3 \pmod{77}$, which is congruent to $64 \cdot 15^4 \pmod{77}$. Since $15^2 \equiv -6 \pmod{77}$, this is congruent to $64 \cdot 36 \pmod{77}$, which is congruent to $(-13) \cdot 36$, which equals -468. Of course, -468 is congruent to -468 plus any multiple of 77. Now $7 \cdot (77) = 539$. Hence $15^{37} \equiv -468 \equiv -468 + 539 \equiv 71 \pmod{77}$. Therefore, we have decrypted the received message, 15, and obtained the original message, 71. (The number 71 must be the original message, since it is the only natural number less than N that is congruent to 71.)

The above looks somewhat complicated. We now proceed to explain and analyze the method in more detail.

For p and q distinct primes and $N = pq$, we use the notation $\phi(N)$ to denote $(p - 1)(q - 1)$. (This is a particular case of a more general concept, known as the *Euler ϕ function*, that we will introduce in the next chapter.) The theorem that underlies the RSA technique is an easy consequence of Fermat's Theorem (5.1.2).

Theorem 6.1.2. *Let $N = pq$, where p and q are distinct prime numbers, and let $\phi(N) = (p-1)(q-1)$. If k and a are any natural numbers, then $a \cdot a^{k\phi(N)} \equiv a$ (mod N).*

Proof. The conclusion of the theorem is equivalent to the assertion that N divides the product of a and $a^{k(p-1)(q-1)} - 1$. Since N is the product of the distinct primes p and q, this is equivalent to the product being divisible by both p and q. (For if any natural number l is divisible by both p and q, then $l = pr$ for some natural number r. Since q divides l and q does not divide p, it follows from Corollary 4.1.3 that q divides r. Thus, $l = pqs$ for some s.)

Consider p (obviously the same proof works for q). There are two cases. First, if p divides a, then p certainly divides $a \cdot (a^{k(p-1)(q-1)} - 1)$. If p does not divide a, then, by Fermat's Theorem (5.1.2), $a^{p-1} \equiv 1$ (mod p). Raising both sides of this congruence to the power $k(q-1)$ shows that $a^{k(p-1)(q-1)} \equiv 1$ (mod p). Thus, p divides $a^{k(p-1)(q-1)} - 1$, so it also divides $a \cdot \left(a^{k(p-1)(q-1)} - 1\right)$. This establishes the result in the case that p does not divide a. Thus, in both cases, p divides $\left(a \cdot a^{k(p-1)(q-1)} - a\right)$. Therefore, $a \cdot a^{k\phi(N)} \equiv a$ (mod N). $\square$

How does this theorem apply to the RSA method? We pick as an encryptor, E, any natural number that does not have any factor in common with $\phi(N)$. As we shall see in the next chapter, this implies that there is a natural number D such that ED is equal to the sum of 1 and a multiple of $\phi(N)$; that is, there is a D such that $ED = 1 + k\phi(N)$ for some natural number k. The theorem we have just proven shows that D is a decryptor, as follows. Suppose that M is the original message, so that $R \equiv M^E$ (mod N) is its encryption. Since R is congruent to M^E modulo N, R^D is congruent to M^{ED} modulo N. But $ED = 1 + k\phi(N)$, so R^D is congruent to $M^{1+k\phi(N)}$ modulo N. This is congruent to the product of M and $M^{k(p-1)(q-1)}$, which is congruent to M by the above theorem. (Of course, M is a natural number less than N, which uniquely determines it.)

The explanation of how to find decryptors requires some additional mathematical tools that we develop in the next chapter. If n is very small, decryptors can be found simply by trial and error.

A complete description of the RSA technique, including choosing encryptors and finding decryptors, is given in the next chapter (see The RSA Procedure for Encrypting Messages 7.2.5).

6.2 Problems

Basic Exercises

1. You are to receive a message using the RSA system. You choose $p = 5$, $q = 7$, and $E = 5$. Verify that $D = 5$ is a decryptor. The encrypted message you receive is 17. What is the actual (decrypted) message?
2. Use the RSA system with $N = 21$ and the encryptor $E = 5$.

(a) Encrypt the message $M = 7$.

(b) Verify that $D = 5$ is a decryptor.

(c) Decrypt the encrypted form of the message.

3. A person tries to receive messages without you being able to decrypt them. The person announces $N = 15$ and $E = 7$ to the world; the person uses such low numbers assuming that you don't understand RSA. An encrypted message $R = 8$ is sent. By trial and error, find a decryptor, D, and use it to find the original message.

Chapter 7
The Euclidean Algorithm and Applications

Each pair of natural numbers has a *greatest common divisor*; i.e., a largest natural number that is a factor of both of the numbers in the pair. For example, the greatest common divisor of 27 and 15 is 3, the greatest common divisor of 36 and 48 is 12, the greatest common divisor of 257 and 101 is 1, the greatest common divisor of 4 and 20 is 4, the greatest common divisor of 7 and 7 is 7, and so on.

Notation 7.0.1. The *greatest common divisor* of the natural numbers m and n is denoted $\gcd(m, n)$.

Thus, $\gcd(27, 15) = 3$, $\gcd(36, 48) = 12$, $\gcd(7, 21) = 7$, and so on. One way to find the greatest common divisor of a pair of natural numbers is by factoring the numbers into primes. Then the greatest common divisor of the two numbers is obtained in the following way: for each prime that occurs as a factor of both numbers, find the highest power of that prime that is a common factor of both numbers and then multiply all those primes to all those powers together to get the greatest common divisor. For example, since $48 = 2^4 \cdot 3$ and $56 = 2^3 \cdot 7$, $\gcd(24, 56) = 2^3 = 8$. As another example, note that $\gcd(1292, 14440) = 76$, since $1292 = 2^2 \cdot 17 \cdot 19$ and $14440 = 2^3 \cdot 5 \cdot 19^2$ and $2^2 \cdot 19 = 76$.

Another way of finding the greatest common divisor of two natural numbers is by using what is called the *Euclidean Algorithm*. One advantage of this method is that it provides a way of expressing the greatest common divisor as a combination of the two original numbers in a way that can be extremely useful. In particular, this technique will allow us to compute a decryptor for each encryptor chosen for RSA coding. As we shall see, other applications of the Euclidean Algorithm include a method for finding integer solutions of linear equations in two variables (*Diophantine equations*) and a different proof of the Fundamental Theorem of Arithmetic.

D. Rosenthal et al., *A Readable Introduction to Real Mathematics*,
Undergraduate Texts in Mathematics, DOI 10.1007/978-3-319-05654-8_7,
© Springer International Publishing Switzerland 2014

7.1   The Euclidean Algorithm

The Euclidean Algorithm is based on the ordinary operation of division of natural numbers, allowing for a remainder. We can express that concept of division as follows (the term "nonnegative integer" refers to either a natural number or 0): if a and b are any natural numbers, then there exist nonnegative integers q and r such that $a = bq + r$ and $0 \leq r < b$. (The number q is called the *quotient* and the number r is called the *remainder* in this equation.) If b divides a, then, of course, $r = 0$.

Let a and b be natural numbers. The Euclidean Algorithm for finding the greatest common divisor of a and b is the following technique. If $b = a$, then clearly the greatest common divisor is a. Suppose that b is less than a. (If b is greater than a, interchange the roles of a and b in the following proof.) Divide a by b as described above to get q and r satisfying $a = bq + r$ with $0 \leq r < b$. If $r = 0$, then clearly the greatest common divisor of a and b is b itself. If r is not 0, divide r into b, to get $b = rq_1 + r_1$, where $0 \leq r_1 < r$. If $r_1 = 0$, stop here. If r_1 is different from 0, divide r_1 into r to get $r = r_1q_2 + r_2$, where $0 \leq r_2 < r_1$. Continue this process until there is the remainder 0. (That will have to occur eventually since the remainders are all nonnegative integers and each one is less than the preceding one.) Thus, there is a sequence of equations as follows:

$$a = bq + r$$
$$b = rq_1 + r_1$$
$$r = r_1q_2 + r_2$$
$$r_1 = r_2q_3 + r_3$$
$$\vdots$$
$$r_{k-3} = r_{k-2}q_{k-1} + r_{k-1}$$
$$r_{k-2} = r_{k-1}q_k + r_k$$
$$r_{k-1} = r_kq_{k+1}$$

It follows that r_k is the greatest common divisor of the original a and b. To see this, note first that r_k is a common divisor of a and b. This can be seen by "working your way up" the equations. Replacing r_{k-1} by r_kq_{k+1} in the next to last equation gives $r_{k-2} = r_kq_{k+1}q_k + r_k = r_k(q_{k+1}q_k + 1)$. Thus, r_k divides r_{k-2}. The equation for r_{k-3} can then be rewritten:

$$r_{k-3} = r_k(q_{k+1}q_k + 1)q_{k-1} + r_kq_{k+1} = r_k\big((q_{k+1}q_k + 1)q_{k-1} + q_{k+1}\big)$$

Thus, r_{k-3} is also divisible by r_k. Continuing to work upwards eventually shows that r_k divides r, then b, and then a. Therefore, r_k is a common divisor of a and b.

To show that r_k is the greatest common divisor of a and b, we show that every other common divisor of a and b divides r_k. Suppose, then, that d is a natural number that divides both a and b. The equation $a = bq + r$ shows that d also divides r. Since d divides both b and r, it divides r_1; since it divides r and r_1, it divides r_2; and so on. Eventually, we see that d also divides r_k. Hence, every common divisor of a and b divides r_k, so r_k is the greatest common divisor of a and b.

Let's look at an example. Suppose we want to use the Euclidean Algorithm to find the greatest common divisor of 33 and 24. We begin with $33 = 24 \cdot 1 + 9$. Then, $24 = 9 \cdot 2 + 6$. Then, $9 = 6 \cdot 1 + 3$. Then, $6 = 3 \cdot 2$. Thus, the greatest common divisor of 33 and 24 is 3.

Definition 7.1.1. We say that the integer d is a *linear combination of the integers a and b* if there exist integers x and y such that $ax + by = d$.

Obtaining the greatest common divisor by the Euclidean Algorithm allows us to express the greatest common divisor as a linear combination of the original numbers, as follows. First consider the above example. From the next to last equation, we get $3 = 9 - 6 \cdot 1$. Substituting the expression for 6 obtained from the previous equation into this one gives

$$3 = 9 - (24 - 9 \cdot 2) \cdot 1 = 9 - 24 + 9 \cdot 2 = 9 \cdot 3 - 24$$

Then solve for 9 in the first equation $9 = 33 - 24 \cdot 1$ and substitute this into the above equation to get $3 = (33 - 24 \cdot 1) \cdot 3 - 24 = 33 \cdot 3 - 24 \cdot 4$. Therefore, $3 = 33 \cdot 3 + 24(-4)$. The greatest common divisor of the numbers 33 and 24, 3, is expressed in the last equation as a linear combination of 33 and 24.

In general, a linear combination of the integers a and b is an expression of the form $ax + by$, where x and y are integers. The Euclidean Algorithm can always be used, as in the above example, to write the greatest common divisor of two natural numbers as a linear combination of those numbers. That is, given natural numbers a and b with greatest common divisor d, there exist integers x and y such that $d = ax + by$. This can be seen by working upwards in the sequence of equations that constitute the Euclidean Algorithm, as in the above example. The next to last equation can be used to write the greatest common divisor, r_k, as a linear combination of r_{k-1} and r_{k-2}; simply solve the next to last equation for r_k. Solving for r_{k-1} in the previous equation and substituting represents r_k as a linear combination of r_{k-2} and r_{k-3}. By continuing to work our way up the ladder of equations in the Euclidean Algorithm, we eventually obtain r_k as a linear combination of the given numbers a and b.

7.2 Applications

Definition 7.2.1. The integers m and n are said to be *relatively prime* if their only common divisor is 1; that is, if $\gcd(m, n) = 1$.

By the above-described consequence of the Euclidean Algorithm, $\gcd(m, n) = 1$ implies that there exist integers s and t such that $sm + tn = 1$. This fact forms the basis for a different proof of the Fundamental Theorem of Arithmetic (4.1.1). We begin by using this fact to prove the following lemma. (You will notice that this is a restatement of Corollary 4.1.3; however, it is presented with a new and independent proof.)

Lemma 7.2.2. *If a prime number divides the product of two natural numbers, then it divides at least one of the numbers.*

Proof. Suppose that p is prime and p divides ab. If p divides a, then we are done. So suppose that p does not divide a; we show that in this case p divides b. Since p is prime, the only possible factors that a could have in common with p are 1 and p. Therefore, a and p are relatively prime and so there exist integers x and y such that $ax + py = 1$. Multiply through by b, getting $bax + bpy = b$. Since p divides ab, it divides the left side of this equation, so it must divide b.   □

We need a slightly stronger lemma which follows easily.

Lemma 7.2.3. *For any natural number n, if a prime divides the product of n natural numbers, then it divides at least one of the numbers.*

Proof. This is a simple consequence of the previous lemma and mathematical induction. The previous lemma is the case $n = 2$. Suppose that the result is true for n factors, where n is greater than 2. Suppose that p is prime and p divides $a_1 a_2 \cdots a_{n+1}$. If p does not divide a_1, then by the case $n = 2$, p divides $a_2 \cdots a_{n+1}$. Hence, by the inductive hypothesis, p divides at least one of $a_2, a_3, \ldots, a_{n+1}$.   □

We are now able to present another proof of the Fundamental Theorem of Arithmetic.

Theorem 7.2.4. *The factorization of a natural number greater than 1 into primes is unique except for the order of the primes.*

Proof. If there were natural numbers with two distinct factorizations, then, by the Well-Ordering Principle (2.1.2), there would exist a smallest such natural number, say N. Then $N = p_1^{\alpha_1} p_2^{\alpha_2} \cdots p_k^{\alpha_k} = q_1^{\beta_1} q_2^{\beta_2} \cdots q_l^{\beta_l}$. Since p_1 divides N, it divides $q_1^{\beta_1} q_2^{\beta_2} \cdots q_l^{\beta_l}$. By Lemma 7.2.3, p_1 divides some q_j. Since p_1 and q_j are prime, $p_1 = q_j$. Dividing both expressions for N by this common factor would then yield a smaller natural number with two distinct factorizations. This contradiction establishes the result.   □

By the consequence of the Euclidean Algorithm noted above, $\gcd(m, n) = 1$ implies that there exist integers s and t such that $sm + tn = 1$. It follows that $sm = 1 - tn$. If m, n, and s are all positive, this equation clearly implies that t is negative. These facts are exactly what we need in order to find decryptors in the RSA system.

Before explaining this in general, let's illustrate it in the case of Example 6.1.1 from the previous chapter. In that example, we started with $p = 7$ and $q = 11$, so that $N = 77$ and $\phi(N) = 6 \cdot 10 = 60$. We took the encryptor $E = 13$. The crucial property of the encryptor is that it is relatively prime to $\phi(N)$. That is true in this case; clearly the only common factor of 13 and 60 is 1. Since $\gcd(13, 60) = 1$, the consequence of the Euclidean Algorithm discussed above implies that there exist integers s and t such that $1 = 13s + 60t$, or $13s = 1 - 60t$. Note that if s and t satisfy this equation, then, for every m, $13(s + 60m) = 1 - 60(t - 13m)$, since this latter equation is obtained from the previous one by simply adding $13 \cdot 60m$ to both sides of the equation. Thus, if the original s was negative, we could choose a positive m large enough so that $s + 60m$ is positive. Therefore, without loss of generality, we can assume that s is positive, which forces t to be negative in the equation $13s = 1 - 60t$. Replace $-t$ by u; then $13s = 1 + 60u$, with s and u both positive integers. We will find such s and u using the Euclidean Algorithm. First, however, note that any such s is a decryptor. To see this, first note that, as in Example 6.1.1, M^{13} is congruent to the encrypted version of the message M. Thus, the encrypted version of the message to the power s is congruent modulo 77 to $(M^{13})^s = M^{13s} = M^{1+60u} = M \cdot M^{60u}$, which is congruent modulo 77 to M by Theorem 6.1.2.

To obtain a decryptor for this example, we begin by using the Euclidean Algorithm to find $\gcd(13, 60)$:

$$60 = 13 \cdot 4 + 8$$
$$13 = 8 \cdot 1 + 5$$
$$8 = 5 \cdot 1 + 3$$
$$5 = 3 \cdot 1 + 2$$
$$3 = 2 \cdot 1 + 1$$
$$2 = 1 \cdot 2$$

Thus, the greatest common divisor of 13 and 60 is 1. Of course, we knew that already; we chose 13 to be relatively prime to 60. The point of using the Euclidean Algorithm is that it allows us to express 1 as a linear combination of 13 and 60, as follows. From the above equation $3 = 2 \cdot 1 + 1$ we get $1 = 3 - 2$. Since $5 = 3 \cdot 1 + 2$, we have $1 = 3 - 2 = 3 - (5 - 3) = 2 \cdot 3 - 5$.

Continuing by working our way up and collecting coefficients gives the following:

$$1 = 2 \cdot 3 - 5$$
$$= 2 \cdot (8 - 5) - 5$$
$$= 2 \cdot 8 - 3 \cdot 5$$
$$= 2 \cdot 8 - 3 \cdot (13 - 8)$$
$$= 5 \cdot 8 - 3 \cdot 13$$
$$= 5 \cdot (60 - 4 \cdot 13) - 3 \cdot 13$$
$$= 5 \cdot 60 - 23 \cdot 13$$

Equivalently, $1 - 5 \cdot 60 = -(23 \cdot 13)$. We are not done. We must find positive integers k and D such that $1 + k \cdot 60 = 13D$. For any integer m, adding $-13 \cdot 60m$ to both sides of the above equation gives $1 - (5 + 13m) \cdot 60 = (-23 - 60m) \cdot 13$. Taking $m = -1$ in this equation gives $1 + 8 \cdot 60 = 37 \cdot 13$. Thus, 37 is a decryptor.

We have illustrated and proven the RSA technique. The following is a statement of what we have established.

The RSA Procedure for Encrypting Messages 7.2.5. *The   recipient   chooses (very large) distinct prime numbers p and q and lets $N = pq$ and $\phi(N) = (p-1)(q-1)$. The recipient then chooses a natural number E (which we are calling the "encryptor" and is often called the "public exponent") greater than 1 that is relatively prime to $\phi(N)$. The pair of numbers (N, E) is called the "public key." The recipient announces the public key and states that any message M consisting of a natural number less than N can be sent as follows: Compute the natural number R less than N such that $M^E \equiv R$ (mod N). The encrypted message that is sent is the natural number R. The recipient decrypts the message by using the Euclidean Algorithm to find natural numbers D (which we are calling the "decryptor" and is often called the "private exponent") and k such that $1 + k\phi(N) = ED$. The pair of numbers (N, D) is called the "private key"; the recipient keeps D secret. The recipient then recovers the original message M as the natural number less than N that is congruent to M^{ED} (mod N).*

The technique that we used to find decryptors can be used to solve many other practical problems.

Definition 7.2.6. A *linear Diophantine equation* is an equation of the form $ax + by = c$, where a, b, and c are integers, and for which we seek solutions (x, y), where x and y are integers.

Example 7.2.7. A store sells two different kinds of boxes of candies. One kind sells for 9 dollars a box and the other kind for 16 dollars a box. At the end of the day, the store has received 143 dollars from the sale of boxes of candy. How many boxes did the store sell at each price?

How can we approach this problem? If x is the number of the less expensive boxes sold and y is the number of the more expensive boxes sold, then the information we are given is

$$9x + 16y = 143$$

There are obviously an infinite number of pairs (x, y) of real numbers that satisfy this equation; the graph in the plane of the set of solutions is a straight line. However, we know more about x and y than simply that they satisfy that equation. We also know that they must both be nonnegative integers. Are there nonnegative integral solutions? Are there any integral solutions at all? Since 9 and 16 are relatively prime, the Euclidean Algorithm tells us that there exist integers s and t (possibly negative) satisfying $9s + 16t = 1$. Multiplying through by 143 gives $9(143s) + 16(143t) = 143$. Therefore, there are integral solutions. However, it is not immediately clear whether there are nonnegative integral solutions, which the actual problem requires. Let's investigate.

We will use the Euclidean Algorithm to find integral solutions to the equation $9s + 16t = 1$. We first use the Euclidean Algorithm to find the greatest common divisor (even though we know it already):

$$16 = 9 \cdot 1 + 7$$
$$9 = 7 \cdot 1 + 2$$
$$7 = 2 \cdot 3 + 1$$
$$2 = 1 \cdot 2$$

Working our way back upwards to express 1 as a linear combination of 9 and 16 gives

$$1 = 7 - 3 \cdot 2$$
$$= 7 - 3 \cdot (9 - 7)$$
$$= 4 \cdot 7 - 3 \cdot 9$$
$$= 4 \cdot (16 - 9) - 3 \cdot 9$$
$$= 16 \cdot 4 - 9 \cdot 7$$

Therefore, $9(-7) + 16 \cdot 4 = 1$. Multiplying by 143 yields $9(-7 \cdot 143) + 16(4 \cdot 143) = 143$. Note that $7 \cdot 143 = 1001$ and $4 \cdot 143 = 572$. For any integer m, we can add and subtract $16 \cdot 9m$; thus, for every integer m,

$$9(-1001 - 16m) + 16(572 + 9m) = 143$$

This gives infinitely many integer solutions; what about nonnegative solutions?

We require that $-1001 - 16m$ be at least 0. That is equivalent to $16m \leq -1001$, or $m \leq \frac{-1001}{16}$. Thus, $m \leq -62.5625$. The largest m satisfying this inequality is $m = -63$. When $m = -63$, $-1001 - 16m = 7$ and $572 + 9m = 5$. Thus, one pair of nonnegative solutions to the original equation is $x = 7$ and $y = 5$. Are there other nonnegative solutions? We will show that all the solutions of this equation are of the form $x = -1001 - 16m$ and $y = 572 + 9m$, for some integer m (see Example 7.2.11 below). To show that the only nonnegative solution is $(7, 5)$ we reason as follows. If we take the next largest m, $m = -64$, then the y we get is $572 - 9 \cdot 64 = -4$. Obviously, if m is even smaller, $572 + 9m$ will be even more negative. Therefore, the only pair of nonnegative solutions to the original equation is $(7, 5)$. Thus, the store sold 7 of the cheaper boxes and 5 of the more expensive boxes of candy.

The basic theorem about solutions of linear Diophantine equations is the following.

Theorem 7.2.8. *The Diophantine equation $ax + by = c$, with a, b, and c integers, has integral solutions if and only if $\gcd(a, b)$ divides c.*

Proof. Let $d = \gcd(a, b)$. If there is a pair of integers (x, y) satisfying the equation, then $ax + by = c$ and, since d divides both of a and b, it follows that d divides c. This proves the easy part of the theorem.

The converse is also easy, but only because of what we learned about the Euclidean Algorithm. We used the Euclidean Algorithm to prove that there exists a pair (s, t) of integers satisfying $as + bt = d$. If d divides c, then there is a k satisfying $c = dk$. Let $x = sk$ and $y = tk$. Then clearly $ax + by = c$.   □

As we've seen in the example where we determined the number of boxes of each kind of candy sold (Example 7.2.7), it is sometimes important to be able to determine all the solutions of a Diophantine equation. In both the decryptor and the candy examples above, we use the very easy fact that $(x + bm, y - am)$ is a solution of $ax + by = c$ whenever (x, y) is a solution. (This follows since $a(x + bm) + b(y - am) = ax + abm + by - abm = ax + by.$) This shows that a Diophantine equation has an infinite number of solutions if it has any solution at all. In finding decryptors, we don't care if the decryptor that we find is only one of a number of possible decryptors. However, in other situations, such as the problem about determining the number of different kinds of boxes of candy that were sold, it is important to have a unique solution that satisfies some other condition of the problem (such as requiring that both of x and y be nonnegative). Theorem 7.2.10 below precisely describes all the solutions of a given linear Diophantine equation.

We require a lemma that generalizes the fact that if a prime divides a product, then it divides at least one of the factors (Lemma 7.2.2).

Lemma 7.2.9. *If s divides tu and s is relatively prime to u, then s divides t.*

Proof. The hypothesis implies that there exists an r such that $tu = rs$. Write the canonical factorization of u into primes, $u = p_1^{\alpha_1} p_2^{\alpha_2} \cdots p_k^{\alpha_k}$. Then,

$$t p_1^{\alpha_1} p_2^{\alpha_2} \cdots p_k^{\alpha_k} = rs$$

Imagine factoring both sides of this equation into a product of primes. By the Fundamental Theorem of Arithmetic (see 4.1.1 or 7.2.4), the factorization of the left-hand side into primes has to be the same as the factorization of the right-hand side. Since s is relatively prime to u, none of the primes comprising s are among the p_i. Thus, all the primes in s occur to at least the same power in the factorization of t, and thus, s divides t. This proves the lemma.   □

Theorem 7.2.10. *Let $\gcd(a, b) = d$. The linear Diophantine equation $ax + by = c$ has a solution if and only if d divides c. If d does divide c and (x_0, y_0) is a solution, then the integral solutions of the equation consist of all the pairs $\left(x_0 + m \cdot \frac{b}{d}, y_0 - m \cdot \frac{a}{d}\right)$, where m assumes all integral values.*

Proof. We already established the first assertion, the criterion for the existence of a solution (Theorem 7.2.8). If (x_0, y_0) is a solution, it is easy to see that each of the other pairs is also a solution, for

$$a\left(x_0 + m \cdot \frac{b}{d}\right) + b\left(y_0 - m \cdot \frac{a}{d}\right) = ax_0 + m \cdot \frac{ab}{d} + by_0 - m \cdot \frac{ab}{d} = ax_0 + by_0 = c$$

All that remains to be proven is that there are no solutions other than those described in the theorem. To see this, suppose that (x_0, y_0) is a solution and that (x, y) is any other solution of $ax + by = c$. Since $ax_0 + by_0 = c$, we can subtract the first equation from the second to conclude that

$$a(x - x_0) + b(y - y_0) = 0$$

Bring one of the terms to the other side and divide both sides of this equation by d to get

$$\frac{a}{d}(x - x_0) = \frac{b}{d}(y_0 - y)$$

Note that $\frac{a}{d}$ and $\frac{b}{d}$ are relatively prime. (For if e was a common factor greater than 1, then $d \cdot e$ would be a common divisor of a and b greater than d.) Hence, by Lemma 7.2.9, $\frac{a}{d}$ divides $(y_0 - y)$ and $\frac{b}{d}$ divides $(x - x_0)$. That is, there are integers k and l such that $y_0 - y = k \cdot \frac{a}{d}$ and $x - x_0 = l \cdot \frac{b}{d}$. Equivalently, $y = y_0 - k \cdot \frac{a}{d}$ and $x = x_0 + l \cdot \frac{b}{d}$. For (x, y) to be a solution, we must have

$$a\left(x_0 + l \cdot \frac{b}{d}\right) + b\left(y_0 - k \cdot \frac{a}{d}\right) = c$$

Thus,

$$ax_0 + l \cdot \frac{ab}{d} + by_0 - k \cdot \frac{ba}{d} = c$$

Since $ax_0 + by_0 = c$, we get $l \cdot \frac{ab}{d} - k \cdot \frac{ba}{d} = 0$. Thus, $l = k$. Call this common value m. Then,

$$x = x_0 + m \cdot \frac{b}{d}$$

$$y = y_0 - m \cdot \frac{a}{d}$$

This proves the theorem. $\qquad\square$

Example 7.2.11. The uniqueness of the solution to the "candy boxes problem" (Example 7.2.7) follows from this theorem. In that example, $\gcd(9, 16) = 1$, so all the solutions are indeed of the form $(-1001 - 16m, 572 + 9m)$.

There are many other interesting applications of the theorem concerning solutions of linear Diophantine equations (see, for example, the problems at the end of this chapter).

Recall that we used the notation $\phi(N)$ to denote $(p-1)(q-1)$ when we were describing the RSA technique with $N = pq$, where p and q were distinct prime numbers. This is a special case of notation for a useful general concept.

Definition 7.2.12. For any natural number m, the *Euler ϕ function*, $\phi(m)$, is defined to be the number of numbers in $\{1, 2, \ldots, m-1\}$ that are relatively prime to m.

Example 7.2.13. To compute $\phi(8)$, we consider the set $\{1, 2, 3, 4, 5, 6, 7\}$. We get $\phi(8) = 4$, since 1, 3, 5, and 7 are the numbers in the set that are relatively prime to 8. Similarly, $\phi(7) = 6$, and $\phi(12) = 4$.

Theorem 7.2.14. *If p is prime, then $\phi(p) = p - 1$.*

Proof. Since p is prime, every number in $\{1, 2, \ldots, p-1\}$ is relatively prime to p, so $\phi(p) = p - 1$.   □

In discussing the RSA technique, we used the notation $\phi(pq) = (p-1)(q-1)$ when p and q were distinct primes. This is consistent with the definition of ϕ we are now using.

Theorem 7.2.15. *If p and q are distinct primes, then $\phi(pq) = (p-1)(q-1)$.*

Proof. Suppose that p and q are primes with p less than q (since they are different, one of them is less than the other), and let $N = pq$. Consider the set $S = \{1, 2, 3, \ldots, p, \ldots, q, \ldots, pq - 1\}$. To find $\phi(N)$, we must determine how many numbers in this set are relatively prime to N. If a number is not relatively prime to N, then it must be divisible by either p or q or both. However, an element k of S cannot be divisible by both p and q. For if k is divisible by p, then $k = pl$ for some natural number l. If k is also divisible by q, then q divides l, since q and p are distinct primes (by Lemma 7.2.2). Thus, $k = pqm$ for some natural number m. It follows that k is at least as big as pq.

There are a total of $pq - 1$ numbers in S; how many multiples of p are there in S? There is $p, 2p, 3p$, and so on, up to $(q-1)p$, since qp is not in S. Thus, there are $q - 1$ multiples of p in S. Similarly, there are $p - 1$ multiples of q in S. Therefore, there is a total of $(q-1) + (p-1) = p + q - 2$ numbers in S that are not relatively prime to N. Since there are $pq - 1$ numbers in S, the number of numbers in S that are relatively prime to N is

$$\phi(N) = pq - 1 - (p + q - 2) = pq - p - q + 1$$

But $pq - p - q + 1 = (p-1)(q-1)$. Therefore, $\phi(N) = (p-1)(q-1)$.   □

There is a formula for $\phi(m)$ for any natural number m greater than 1, in terms of the canonical factorization of m into a product of primes (see Problem 27 at the end of this chapter).

Fermat's beautiful theorem that $a^{p-1} \equiv 1 \pmod{p}$ (5.1.2) (for primes p and natural numbers a that are not divisible by p) can be generalized to composite moduli. We require the following lemma that generalizes Theorem 5.1.1.

Lemma 7.2.16. *If a is relatively prime to m and $ax \equiv ay \pmod{m}$, then $x \equiv y \pmod{m}$.*

Proof. We are given that m divides $ax - ay$. That is, m divides $a(x - y)$. By Lemma 7.2.9, m divides $x - y$. Thus, $x \equiv y \pmod{m}$.   □

Euler's Theorem 7.2.17. *If m is a natural number greater than 1 and a is a natural number that is relatively prime to m, then $a^{\phi(m)} \equiv 1 \pmod{m}$.*

Proof. The proof is very similar to the proof of Fermat's Theorem (5.1.2). Let $S = \{r_1, r_2, \ldots, r_{\phi(m)}\}$ be the set of numbers in $\{1, 2, \ldots, m-1\}$ that are relatively prime to m. Then let $T = \{ar_1, ar_2, \ldots, ar_{\phi(m)}\}$. Clearly, no two of the numbers in S are congruent to each other, since they are distinct numbers which are all less than m. Note also that no two of the numbers in T are congruent to each other, since $ar_i \equiv ar_j \pmod{m}$ would imply, by Lemma 7.2.16, that $r_i \equiv r_j \pmod{m}$. Moreover, each ar_i is relatively prime to m and therefore so is any number that ar_i is congruent to. Thus, the numbers in $\{ar_1, ar_2, \ldots, ar_{\phi(m)}\}$ are congruent, in some order, to the numbers in $\{r_1, r_2, \ldots, r_{\phi(m)}\}$. It follows, as in the proof of Fermat's Theorem, that the product of all the numbers in T is congruent to the product of all the numbers in S. That is,

$$a \cdot r_1 \cdot a \cdot r_2 \cdots a \cdot r_{\phi(m)} \equiv r_1 r_2 \cdots r_{\phi(m)} \pmod{m}$$

Since $r_1 r_2 \cdots r_{\phi(m)}$ is relatively prime to m, we can divide both sides of this congruence by that product (see Lemma 7.2.16), to get $a^{\phi(m)} \equiv 1 \pmod{m}$.   □

Fermat's Theorem is a special case of Euler's.

Corollary 7.2.18 (Fermat's Theorem). *If p is a prime and p does not divide a, then $a^{p-1} \equiv 1 \pmod{p}$.*

Proof. Since p is prime, the fact that p does not divide a means that a and p are relatively prime. Also, $\phi(p) = p-1$. Thus, Fermat's Theorem follows from Euler's Theorem (7.2.17).   □

7.3 Problems

Basic Exercises

1. Find the greatest common divisor of each of the following pairs of integers in two
   different ways, by using the Euclidean Algorithm and by factoring both numbers
   into primes:

   (a) 252 and 198
   (b) 291 and 573
   (c) 1800 and 240
   (d) 52 and 135

2. For each of the pairs in Problem 1 above, write the greatest common divisor as a
   linear combination of the given numbers.
3. Find integers x and y such that $3x - 98y = 12$.
4. (a) Find a formula for all integer solutions of the Diophantine equation
       $3x + 4y = 14$.
   (b) Find all pairs of natural numbers that solve the above equation.
5. Let ϕ be Euler's ϕ function. Find:

   (a) $\phi(12)$ (e) $\phi(97)$
   (b) $\phi(26)$ (f) $\phi(73)$
   (c) $\phi(21)$ (g) $\phi(101 \cdot 37)$
   (d) $\phi(36)$ (h) $\phi(3^{100})$

6. Use the Euclidean Algorithm to find the decryptors in Problems 1, 2, and 3 in
   Chapter 6.

Interesting Problems

7. Use the Euclidean Algorithm (and a calculator) to find the greatest common
   divisor of each of the following pairs of natural numbers:

   (a) 47,295 and 297
   (b) 77,777 and 2,891

8. Find the smallest natural number x such that $24x$ leaves a remainder of 2 upon
   division by 59.
9. A small theater has a student rate of $3 per ticket and a regular rate of $10 per
   ticket. Last night $243 was collected from the sale of tickets. There were more
   than 50 but less than 60 tickets sold. How many student tickets were sold?
10. A liquid comes in 17 liter and 13 liter cans. Someone needs exactly 287 liters of
    the liquid. How many cans of each size should the person buy?

11. Let a, b, and n be natural numbers. Prove that if a^n and b^n are relatively prime, then a and b are relatively prime.

12. Let a, b, m, and n be natural numbers with m and n greater than 1. Assume that m and n are relatively prime. Prove that if $a \equiv b \pmod{m}$ and $a \equiv b \pmod{n}$, then $a \equiv b \pmod{mn}$.

13. Let a and b be natural numbers.

   (a) Suppose there exist integers m and n such that $am + bn = 1$. Prove that a and b are relatively prime.
   (b) Prove that $5a + 2$ and $7a + 3$ are relatively prime for every natural number a.

14. Let p be a prime number. Prove that $\phi(p^2) = p^2 - p$.

15. The public key $N = 55$ and $E = 7$ is announced. The encrypted message 5 is received.

   (a) Find a decryptor, D, and prove that D is a decryptor.
   (b) Decrypt 5 to find the original message.

16. Find a multiplicative inverse of 2^{29} modulo 9.

17. Prove that a has a multiplicative inverse modulo m if and only if a and m are relatively prime.

Challenging Problems

18. Suppose that a and b are relatively prime natural numbers such that ab is a perfect square. Show that a and b are each perfect squares.

19. Show that if m and n are relatively prime and a and b are any integers, then there is an integer x that simultaneously satisfies the two congruences $x \equiv a \pmod{m}$ and $x \equiv b \pmod{n}$.

20. Generalize the previous problem as follows (this result is called the *Chinese Remainder Theorem*):
   If $\{m_1, m_2, \ldots, m_k\}$ is a collection of natural numbers greater than 1, each pair of which is relatively prime, and if $\{a_1, a_2, \ldots, a_k\}$ is any collection of integers, then there is an integer x that simultaneously satisfies all of the congruences $x \equiv a_j \pmod{m_j}$. Moreover, if x_1 and x_2 are both simultaneous solutions of all of those congruences, then $x_1 \equiv x_2 \pmod{m_1 m_2 \cdots m_k}$.

21. Let p be an odd prime and let $m = 2p$. Prove that $a^{m-1} \equiv a \pmod{m}$ for all natural numbers a.

22. Let a and b be relatively prime natural numbers greater than or equal to 2. Prove that $a^{\phi(b)} + b^{\phi(a)} \equiv 1 \pmod{ab}$.

23. Suppose that a, b, and c are each natural numbers. Prove that there are at most a finite number of pairs of natural numbers (x, y) that satisfy $ax + by = c$.

24. Show that m is prime if there is an integer a such that $a^{m-1} \equiv 1 \pmod{m}$ and $a^k \not\equiv 1 \pmod{m}$ for every natural number $k < m - 1$.

25. Suppose that a and m are relatively prime and that k is the smallest natural number such that a^k is congruent to 1 modulo m. Prove that k divides $\phi(m)$.

26. For p a prime and k a natural number, show that $\phi(p^k) = p^k - p^{k-1}$.

27. If the canonical factorization of the natural number n into primes is $n = p_1^{k_1} \cdot p_2^{k_2} \cdots p_m^{k_m}$, prove that

$$\phi(n) = \left(p_1^{k_1} - p_1^{k_1-1}\right) \cdot \left(p_2^{k_2} - p_2^{k_2-1}\right) \cdots \left(p_m^{k_m} - p_m^{k_m-1}\right)$$

Chapter 8
Rational Numbers and Irrational Numbers

So far, the only numbers that we have been discussing are the "whole numbers;" that is, the integers. There are many other interesting things that can be said about the integers, but, for now, we will move on to consider other numbers, the *rational numbers*, also known as "fractions," and then the *real numbers*.

8.1 Rational Numbers

Definition 8.1.1. A *rational number* is a number of the form $\frac{m}{n}$, where m and n are integers and $n \neq 0$.

Some examples of rational numbers are $\frac{3}{4}, \frac{-7}{23}, \frac{12}{-36}, \frac{1}{2}$, and $\frac{2}{4}$.

Wait a minute. Are $\frac{1}{2}$ and $\frac{2}{4}$ different rational numbers? They are not; they are two different expressions representing the same number. Similarly $\frac{12}{48} = \frac{1}{4}, \frac{-7}{3} = \frac{7}{-3}$, $\frac{16}{2} = \frac{8}{1}$, and so on. The condition under which two different expressions as quotients of integers represent the same rational number is the following.

Definition 8.1.2. The rational number $\frac{m_1}{n_1}$ is equal to the rational number $\frac{m_2}{n_2}$ when $m_1 n_2 = m_2 n_1$.

Thus, when we use the representation $\frac{1}{2}$, we recognize that we are representing a number that could also be denoted $\frac{2}{4}, \frac{-3}{-6}$, and so on.

Why don't we allow 0 denominators in the expressions for rational numbers? If we did allow 0 denominators, the arithmetic would be very peculiar. For example, $\frac{7}{0}$ would equal $\frac{-12}{0}$, since $7 \cdot 0 = -12 \cdot 0$. In fact, we would have $\frac{a}{0} = \frac{b}{0}$ for all integers a and b. It is not at all useful to have such peculiarities as part of our arithmetic, so we do not allow 0 to be a denominator of any rational number.

Notation 8.1.3. The set of all rational numbers is denoted by $\mathbb{Q}$.

D. Rosenthal et al., *A Readable Introduction to Real Mathematics*,
Undergraduate Texts in Mathematics, DOI 10.1007/978-3-319-05654-8__8,
© Springer International Publishing Switzerland 2014

The operations of multiplication and addition of rational numbers can be defined in terms of the operations on integers.

Definition 8.1.4. The product of the rational numbers $\frac{m_1}{n_1}$ and $\frac{m_2}{n_2}$, denoted $\frac{m_1}{n_1} \cdot \frac{m_2}{n_2}$ or simply $\frac{m_1}{n_1} \frac{m_2}{n_2}$, is the rational number

$$\frac{m_1 m_2}{n_1 n_2}$$

The sum of the rational numbers $\frac{m_1}{n_1}$ and $\frac{m_2}{n_2}$ is the rational number

$$\frac{m_1}{n_1} + \frac{m_2}{n_2} = \frac{m_1 n_2 + m_2 n_1}{n_1 n_2}$$

We can think of the integers as the rational numbers whose denominator is 1; we invariably write them without the denominator. For example, we write -17 for $\frac{-17}{1}$ (and also, of course, for $\frac{-34}{2}$, and so on). In particular, we write 0 for $\frac{0}{1}$ and 1 for $\frac{1}{1}$. Note that from Definition 8.1.4, 0 and 1 are, respectively, additive and multiplicative identities for the rational numbers, as they are for the integers. That is, $\frac{m}{n} + 0 = \frac{m}{n}$ and $\frac{m}{n} \cdot 1 = \frac{m}{n}$, for every rational number $\frac{m}{n}$. Also note that, as is the case with the set of integers, every rational number has an additive inverse: $\frac{m}{n} + \frac{-m}{n} = \frac{0}{n} = 0$.

Definition 8.1.5. A *multiplicative inverse* for the number x is a number y such that $xy = 1$.

Of course, 0 has no multiplicative inverse, since 0 times any number is 0. If x and y are both integers and $xy = 1$, then x and y must both be 1, or -1. Hence, the only integers that have multiplicative inverses within the set of integers are the numbers 1 and -1. In the set of rational numbers, the situation is very different.

Theorem 8.1.6. *If $\frac{m}{n}$ is a rational number other than 0, then $\frac{m}{n}$ has a multiplicative inverse.*

Proof. If $\frac{m}{n} \neq 0$, then $m \neq 0$. Therefore, $\frac{n}{m}$ is also a rational number and $\frac{m}{n} \frac{n}{m} = \frac{mn}{nm} = \frac{1}{1} = 1$. Therefore, $\frac{n}{m}$ is a multiplicative inverse for $\frac{m}{n}$. □

Definition 8.1.7. A *polynomial with integer coefficients* is an expression of the form

$$a_n x^n + a_{n-1} x^{n-1} + \cdots + a_1 x + a_0$$

where n is a nonnegative integer and the a_i are integers with a_n different from 0 (we also include the case where $n = 0$ and $a_0 = 0$). The number x_0 is a *root* (or *zero*) of a polynomial if the value of the polynomial obtained by replacing x by x_0 is 0.

Example 8.1.8. The polynomial $x^5 + x - 1$ has no rational roots.

Proof. Suppose that $\frac{m}{n}$ was a rational root. Without loss of generality we can assume that m and n are relatively prime (if m and n had a common factor, that common factor could be divided out from m and n, getting an equivalent fraction). Substituting $\frac{m}{n}$ in the polynomial would yield $(\frac{m}{n})^5 + \frac{m}{n} - 1 = 0$. Multiplying both sides by n^5 gives $m^5 + mn^4 - n^5 = 0$, or $m(m^4 + n^4) = n^5$. It follows that any prime divisor of m is a divisor of n^5 and, hence, also of n. Since m and n are relatively prime, m has no prime divisors. Thus, m is either 1 or -1. Similarly, the above equation yields $m^5 = n(n^4 - mn^3)$ from which it follows that any prime divisor of m would divide n. Thus, n does not have any prime divisors, so n is either 1 or -1. Therefore, the only possible values of $\frac{m}{n}$ are 1 or -1. That is, the only possible rational roots of the polynomial are 1 and -1. However, it is clear that neither 1 nor -1 is a root. Thus, the polynomial does not have any rational roots. $\square$

There is a general theorem, whose proof is similar to the above example, that is often useful in determining whether or not polynomials have rational roots and may also be used to find such roots.

The Rational Roots Theorem 8.1.9. *If $\frac{m}{n}$ is a rational root of the polynomial $a_k x^k + a_{k-1} x^{k-1} + \cdots + a_1 x + a_0$, where the a_j are integers and m and n are relatively prime, then m divides a_0 and n divides a_k.*

Proof. Assuming that $\frac{m}{n}$ is a root gives

$$a_k \left(\frac{m}{n}\right)^k + a_{k-1} \left(\frac{m}{n}\right)^{k-1} + \cdots + a_1 \left(\frac{m}{n}\right) + a_0 = 0$$

Multiplying both sides of this equation by n^k produces the equation

$$a_k m^k + a_{k-1} m^{k-1} n + \cdots + a_1 m n^{k-1} + a_0 n^k = 0$$

It follows that

$$m(a_k m^{k-1} + a_{k-1} m^{k-2} n + \cdots + a_1 n^{k-1}) = -a_0 n^k$$

Since m and n are relatively prime, m and n^k are also relatively prime. On the other hand, m divides $-a_0 n^k$. Thus, by Lemma 7.2.9, m divides a_0. Similarly,

$$a_k m^k = -(a_{k-1} m^{k-1} n + \cdots + a_1 m n^{k-1} + a_0 n^k)$$

so,

$$a_k m^k = -n(a_{k-1} m^{k-1} + \cdots + a_1 m n^{k-2} + a_0 n^{k-1})$$

Since n is relatively prime to m and n divides $a_k m^k$, it follows (by Lemma 7.2.9) that n divides a_k. This proves the theorem. $\square$

Example 8.1.10. Find all the rational roots of the polynomial $2x^3 - x^2 + x - 6$.

Proof. By the Rational Roots Theorem (8.1.9), any rational root $\frac{m}{n}$ in lowest terms has the property that n divides 2 and m divides 6. Thus, the only possible values of n are $1, -1, 2, -2$, and the only possible values of m are $6, -6, 3, -3, 2, -2, 1, -1$. The possible values of the quotient $\frac{m}{n}$ are therefore $6, -6, 3, -3, 2, -2, \frac{3}{2}, -\frac{3}{2}, 1, -1, \frac{1}{2}, -\frac{1}{2}$. We can determine which of these possible roots actually are roots by simply substituting them for x and seeing if the result is 0. In this example, the only rational root is $\frac{3}{2}$.   □

8.2 Irrational Numbers

In a sense, all actual computations, by human or electronic computers, are done with rational numbers. However, it is important, within mathematics itself and in using mathematics to understand the world, to have other numbers as well.

Example 8.2.1. Suppose that you walk one mile due east and then one mile due north. How far are you from your starting point? The straight line from your starting point to your final position is the hypotenuse of a right triangle whose legs are each one mile long. The length of the hypotenuse is the distance that you are from your starting point. If x denotes that distance, then the Pythagorean Theorem (11.3.7) tells us that $x^2 = 2$.

It is obviously useful to have *some* number that denotes that distance. Is there a rational number x such that $x^2 = 2$? This question can be rephrased: are there integers m and n with $n \neq 0$ such that $(\frac{m}{n})^2 = 2$? This, of course, is equivalent to the question of whether there are integers m and n different from 0 that satisfy the equation $m^2 = 2n^2$. This is a very concrete question about integers; what is the answer?

Theorem 8.2.2. *There do not exist integers m and n with $n \neq 0$ such that* $(\frac{m}{n})^2 = 2$.

Proof. Suppose that there did exist such m and n. We will show that this assumption leads to a contradiction. From $(\frac{m}{n})^2 = 2$ it would follow that $m^2 = 2n^2$. The equation $m^2 = 2n^2$ implies that m^2 is an even number, since it is the product of 2 and another number. What about m itself? If m was odd, then $m - 1$ would have to be even, so $m - 1 = 2k$ for some integer k, or $m = 2k + 1$. It would follow from this that $m^2 = (2k + 1)^2 = 4k^2 + 4k + 1 = 2(2k^2 + 2k) + 1$, which is an odd number (since it is 1 more than a multiple of 2). Thus, if m was odd, m^2 would have to be odd. Since m^2 is even, we conclude that m is even. Therefore, $m = 2s$ for some integer s, from which it follows that $m^2 = 4s^2$. Substituting $4s^2$ for m^2 into the equation $m^2 = 2n^2$ gives $4s^2 = 2n^2$, or $2s^2 = n^2$. Thus, n^2 is an even number and, reasoning as we did above for m, it follows that n itself is an even number.

What have we proven so far? We have proven that $m^2 = 2n^2$ implies that both m and n are even. But if $\frac{m}{n}$ is any rational number with m and n both even integers, then the common factor of 2 can be "divided out" from both m and n, which gives an expression of the number with numerator and denominator each half of the corresponding part of the original representation of the number. This process of dividing out 2's can be repeated until at least one of the numerator and denominator is odd. That is, there exists m_0 and n_0 different from 0 such that at least one of m_0 and n_0 is odd, and $\frac{m}{n} = \frac{m_0}{n_0}$. Then, $(\frac{m_0}{n_0})^2 = 2$, so the above reasoning would imply that both of m_0 and n_0 are even, which contradicts the fact that at least one of them is odd. $\qquad\square$

We have proven that there is no rational number that satisfies the equation $x^2 = 2$. Is there any number that satisfies this equation? It would obviously be useful to have such a number, for the purpose of specifying how far a person in Example 8.2.1 is from their starting point and for many other purposes. Mathematicians have developed what are called the *real numbers*; the real numbers include numbers for every possible distance. The real numbers can be put into correspondence with the points on a line by labeling one point "0" and marking points to the right of 0 with the distances that they are from 0 (using any fixed units). Points on the line to the left of 0 are labeled with corresponding negative real numbers. The resulting *real number line* looks like

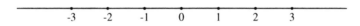

The set of real numbers and the arithmetical operations on them can be precisely constructed in terms of rational numbers. In fact, there are several ways to do that. None of the ways of constructing the real numbers in terms of the rational numbers are easy; they all require substantial development. There are two main approaches, one using *Cauchy sequences* and the other using *Dedekind cuts*. The *Dedekind cuts* approach is outlined in Problem 15 at the end of this chapter. For the present, we will simply assume that the real numbers exist and that the arithmetical operations on them have the usual properties.

Notation 8.2.3. The set of all real numbers is denoted by $\mathbb{R}$.

It can be shown that there is a positive real number x such that $x^2 = 2$. This number is denoted $\sqrt{2}$ or $2^{\frac{1}{2}}$.

Definition 8.2.4. A real number that is not a rational number is said to be *irrational*.

Theorem 8.2.2 shows that $\sqrt{2}$ is not a rational number and thus can be rephrased as follows.

Theorem 8.2.5. *The number $\sqrt{2}$ is irrational.*

The symbol $\sqrt{3}$ represents the positive real number satisfying $(\sqrt{3})^2 = 3$; is $\sqrt{3}$ irrational too?

We can establish a more general result.

Theorem 8.2.6. *If p is a prime number, then $\sqrt{p}$ is irrational.*

Proof. The proof will be similar to that of the special case $p = 2$. Suppose that $n \neq 0$ and m and n are integers satisfying $(\frac{m}{n})^2 = p$. Then $m^2 = pn^2$. Since $m^2 = pn^2$, p divides m^2. Thus, p divides the product $m \cdot m$, from which it follows that p divides at least one of the factors (see Corollary 4.1.3); that is, p divides m. Therefore, there is an integer s such that $m = ps$, which gives $(ps)^2 = pn^2$. Dividing both sides of this equation by p gives $ps^2 = n^2$. Thus, p divides the product $n \cdot n$ and we conclude that p divides n. This shows that whenever $\frac{m}{n}$ is a rational number with $(\frac{m}{n})^2 = p$, both m and n are divisible by p. As in the case where $p = 2$ (see Theorem 8.2.2), the fact that common factors of numerators and denominators of fractions can be "divided out" leads to a contradiction. $\square$

Of course, some natural numbers do have rational square roots. For example, $\sqrt{1} = 1$, $\sqrt{4} = 2$, and $\sqrt{289} = 17$. What about $\sqrt{6}$? More generally is there a natural number m such that $\sqrt{m}$ is rational but $\sqrt{m}$ is not an integer? To answer this question it is useful to begin with the following.

Lemma 8.2.7. *A natural number other than 1 is a perfect square (i.e., the square of a natural number) if and only if every prime number in its canonical factorization occurs to an even power.*

Proof. Let n be a natural number. If the canonical factorization of n (see Corollary 4.1.2) is $n = p_1^{\alpha_1} p_2^{\alpha_2} \cdots p_k^{\alpha_k}$, then $n^2 = p_1^{2\alpha_1} p_2^{2\alpha_2} \cdots p_k^{2\alpha_k}$. The uniqueness of the factorization into primes implies that this expression is the canonical factorization of n^2. All the exponents are obviously even. This proves that the square of every natural number has the property that every exponent in its canonical factorization is even. The converse is even easier. For if $m = p_1^{2\alpha_1} p_2^{2\alpha_2} \cdots p_k^{2\alpha_k}$, then obviously $m = n^2$, where $n = p_1^{\alpha_1} p_2^{\alpha_2} \cdots p_k^{\alpha_k}$. $\square$

Theorem 8.2.8. *If the square root of a natural number is rational, then the square root is an integer.*

Proof. Suppose that N is a natural number and that the square root of N is rational. The case $N = 1$ poses no difficulties. If N is greater than 1, let its canonical factorization be $p_1^{\alpha_1} p_2^{\alpha_2} \cdots p_t^{\alpha_t}$. Since $\sqrt{N}$ is rational, there exist natural numbers m and n such that $\sqrt{N} = \frac{m}{n}$. Let the canonical factorizations of m and n, respectively, be $m = q_1^{\beta_1} q_2^{\beta_2} \cdots q_u^{\beta_u}$ and $n = r_1^{\gamma_1} r_2^{\gamma_2} \cdots r_v^{\gamma_v}$. Since $N = \frac{m^2}{n^2}$, it follows that $n^2 N = m^2$. In terms of the canonical factorizations of N, n, and m, this yields

$$(r_1^{\gamma_1} r_2^{\gamma_2} \cdots r_v^{\gamma_v})^2 p_1^{\alpha_1} p_2^{\alpha_2} \cdots p_t^{\alpha_t} = \left(q_1^{\beta_1} q_2^{\beta_2} \cdots q_u^{\beta_u} \right)^2$$

It follows that

$$r_1^{2\gamma_1} r_2^{2\gamma_2} \cdots r_v^{2\gamma_v} p_1^{\alpha_1} p_2^{\alpha_2} \cdots p_t^{\alpha_t} = q_1^{2\beta_1} q_2^{2\beta_2} \cdots q_u^{2\beta_u}$$

We want to prove that each of the α_i is even. By the uniqueness of the factorization into primes, each p_i is one of the q_j, for some j. Since $2\beta_j$ is even, p_i occurs to an even power on both sides of the equation. Of course, p_i could be one of the r's. If so, since the powers of all the r's are even, the total power that p_i occurs to on the left-hand side of the equation is the sum of α_i and an even number. Since this sum must be the even number $2\beta_j$, it follows that α_i is even. Thus, every α_i is even and the above lemma (8.2.7) implies that N is the square of an integer.   □

Example 8.2.9.   The number $\sqrt[3]{4}$ is irrational.

Proof. If $\sqrt[3]{4} = \frac{m}{n}$ with m and n integers, then $4n^3 = m^3$. Write this equation in terms of the canonical factorizations of m and n, getting

$$4\left(p_1^{\alpha_1} p_2^{\alpha_2} \cdots p_r^{\alpha_r}\right)^3 = \left(q_1^{\beta_1} q_2^{\beta_2} \cdots q_s^{\beta_s}\right)^3$$

So,

$$2^2 \cdot p_1^{3\alpha_1} p_2^{3\alpha_2} \cdots p_r^{3\alpha_r} = q_1^{3\beta_1} q_2^{3\beta_2} \cdots q_s^{3\beta_s}$$

The prime 2 must occur to a power that is a multiple of 3, since every prime on the right-hand side of this equation occurs to such a power. On the other hand, 2 occurs on the left-hand side of the equation to a power that is two more than a multiple of 3. The uniqueness of the factorization into primes implies that no such equation is possible.   □

Example 8.2.10.   The number $\sqrt{3} + \sqrt{5}$ is irrational.

Proof. Suppose that $\sqrt{3} + \sqrt{5} = r$, with r a rational number. Then $\sqrt{3} = r - \sqrt{5}$. Squaring both sides of this equation gives

$$3 = (r - \sqrt{5})^2 = r^2 - 2\sqrt{5}r + 5$$

From this it would follow that $2\sqrt{5}r = r^2 + 2$ or $\sqrt{5} = \frac{r^2+2}{2r}$. But r rational implies that $\frac{r^2+2}{2r}$ is rational, which contradicts the fact that $\sqrt{5}$ is irrational (Theorem 8.2.6).   □

The following is a question with an interesting answer: Do there exist two irrational numbers such that one of them to the power of the other is rational? That is, can x^y be rational if x and y are both irrational? A case that appears to be simple is that of $(\sqrt{3})^{\sqrt{2}}$. In fact, however, it is not at all easy to determine whether or not $(\sqrt{3})^{\sqrt{2}}$ is rational. Nonetheless, this example can still be used to prove that the general question has an affirmative answer, as follows. Either $(\sqrt{3})^{\sqrt{2}}$ is rational or it is irrational. If it is rational, it provides an example showing that the answer to the question is affirmative. If $(\sqrt{3})^{\sqrt{2}}$ is irrational, let $x = (\sqrt{3})^{\sqrt{2}}$ and $y = \sqrt{2}$. Then x^y is an irrational number to an irrational power. But

$$x^y = \left((\sqrt{3})^{\sqrt{2}}\right)^{\sqrt{2}} = (\sqrt{3})^{\sqrt{2}\cdot\sqrt{2}} = (\sqrt{3})^2 = 3$$

This gives an affirmative answer in this case as well. In other words, $(\sqrt{3})^{\sqrt{2}}$ answers our original question, whether it itself is rational or irrational. In fact, $(\sqrt{3})^{\sqrt{2}}$ is irrational, as follows from the Gelfond–Schneider Theorem, a theorem that is very difficult to prove.

8.3   Problems

Basic Exercises

1. Use the Rational Roots Theorem (8.1.9) to find all rational roots of each of the following polynomials (some may not have any rational roots at all):

   (a) $x^2 + 5x + 2$
   (b) $2x^3 - 5x^2 + 14x - 35$
   (c) $x^{10} - x + 1$

2. Show that $\sqrt[3]{5}$ is irrational.

3. Show that $\sqrt{\frac{1}{2}}$ is irrational.

4. Must the sum of an irrational number and a rational number be irrational?

5. Must an irrational number to a rational power be irrational?

6. Must the sum of two irrational numbers be irrational?

7. Is $\sqrt[3]{\sqrt{49}+1}$ irrational?

8. If y is irrational and x is any rational number other than 0, show that xy is irrational.

Interesting Problems

9. Determine whether each of the following numbers is rational or irrational and prove that your answer is correct:

   (a) $32^{\frac{2}{3}}$

   (b) $28^{\frac{2}{5}}$

   (c) $\frac{\sqrt[4]{7}}{\sqrt{5}}$

   (d) $\frac{\sqrt{7}}{\sqrt[3]{15}}$

   (e) $\frac{\sqrt{63}}{\sqrt{28}}$

   (f) $\sqrt{\frac{3}{8}}$

   (g) $\sqrt[7]{\frac{8}{9}}$

10. Prove that $\sqrt[3]{3} + \sqrt{11}$ is irrational.

Challenging Problems

11. Prove that the following numbers are irrational:

    (a) $\sqrt{5} + \sqrt{7}$
    (b) $\sqrt[3]{4} + \sqrt{10}$
    (c) $\sqrt[3]{5} + \sqrt{3}$
    (d) $\sqrt{3} + \sqrt{5} + \sqrt{7}$
    (e) $\sqrt{3} - \frac{\sqrt{5}}{17}$

12. Suppose that a and b are odd natural numbers and $a^2 + b^2 = c^2$. Prove that c is irrational.

13. Let k be a natural number. Prove that the kth root of a natural number is rational if and only if the kth root is a natural number.

14. Prove that if a and b are natural numbers and n is a natural number such that $n^{\frac{a}{b}}$ is rational, then $n^{\frac{a}{b}}$ is a natural number.

15. (Very challenging.) In this problem, we outline the *Dedekind cuts* approach to constructing the real numbers. In this approach, real numbers are defined as certain kinds of subsets of the set of rational numbers. The definition is the following. A *real number* is a nonempty proper subset of the set of rational numbers that does not have a greatest element and has the property that if a rational number t is in the set and a rational number s is less than t, then s is in the set. (A "proper subset" is a subset which is not the whole set.)

    (a) If r is a rational number, the representation of r as a real number is as the set of all rational numbers that are less than r. Prove that such a representation is a real number according to the definition given above.

    (b) If $\mathcal{S}$ and $\mathcal{T}$ are real numbers as defined above, then $\mathcal{S} + \mathcal{T}$ is defined to be the set of all $s + t$ with $s \in \mathcal{S}$ and $t \in \mathcal{T}$. Prove that $\mathcal{S} + \mathcal{T}$ is a real number (i.e., has the above properties).

    (c) Prove that addition of real numbers as defined above is commutative; that is, $\mathcal{S} + \mathcal{T} = \mathcal{T} + \mathcal{S}$ for all real numbers $\mathcal{S}$ and $\mathcal{T}$.

    (d) Prove that addition of real numbers as defined above is associative; that is, $(\mathcal{S}_1 + \mathcal{S}_2) + \mathcal{S}_3 = \mathcal{S}_1 + (\mathcal{S}_2 + \mathcal{S}_3)$ for all real numbers $\mathcal{S}_1$, $\mathcal{S}_2$, and $\mathcal{S}_3$.

    (e) If $\mathcal{S}$ is a real number, define $-\mathcal{S}$ to be the set of all rational numbers t such that $-t$ is not in $\mathcal{S}$ and $-t$ is not the smallest rational number that is not in $\mathcal{S}$. Prove that $-\mathcal{S}$ is a real number whenever $\mathcal{S}$ is a real number.

    (f) Let $\mathcal{O}$ denote the real number corresponding to the rational number 0 (i.e., the set of all x in $\mathbb{Q}$ such that x is less than 0). Prove that $\mathcal{S} + \mathcal{O} = \mathcal{S}$, for every real number $\mathcal{S}$.

    (g) Prove that $\mathcal{S} + (-\mathcal{S}) = \mathcal{O}$, for every real number $\mathcal{S}$.

    (h) We say that the real number $\mathcal{S}$ is *positive* if $\mathcal{S}$ contains a rational number that is greater than 0. If $\mathcal{S}$ and $\mathcal{T}$ are positive real numbers, then the product $\mathcal{S}\mathcal{T}$ is defined to be the union of the set of all rational numbers that are less than or equal to 0 together with the set of all rational numbers of the form

st, where s is a positive number in S and t is a positive number in T. Prove that the product of two positive real numbers is a real number (i.e., satisfies the properties of a real number listed above).

(i) If S is a real number, define $|S|$ to be S if S is a positive real number and $-S$ otherwise. Say that a real number is *negative* if it is not positive and is not $\mathcal{O}$. Prove that $|S|$ is positive for all S not equal to $\mathcal{O}$.

(j) If S and T are real numbers, define the product ST to be $-(|S||T|)$ if one is negative and the other is positive, to be $|S||T|$ if both are negative, and to be $\mathcal{O}$ if either is $\mathcal{O}$. Prove that multiplication of real numbers is commutative; that is, $ST = TS$, for all real numbers S and T.

(k) Let $\mathcal{I}$ denote the real number corresponding to the rational number 1 (i.e., the set of all x in $\mathbb{Q}$ such that x is less than 1). Prove that the product of $\mathcal{I}$ and S is S, for every real number S.

(l) For a positive real number S, define $\frac{1}{S}$ to be the union of the set of rational numbers that are less than or equal to 0 and the set of rational numbers t such that $\frac{1}{t}$ is not in S and $\frac{1}{t}$ is not the smallest rational number not in S. Prove that $\frac{1}{S}$ is a real number whenever S is a positive real number.

(m) For S a negative real number, define $\frac{1}{S}$ to be $-\frac{1}{|S|}$. For S any real number other than $\mathcal{O}$, prove that the product of S and $\frac{1}{S}$ is $\mathcal{I}$.

(n) Prove that multiplication of real numbers as defined above is associative; that is, $(S_1 S_2)S_3 = S_1(S_2 S_3)$ for all real numbers S_1, S_2, and S_3.

(o) Prove that multiplication of real numbers is distributive over addition; that is, $S_1(S_2 + S_3) = S_1 S_2 + S_1 S_3$ for all real numbers S_1, S_2, and S_3.

(p) (Existence of the square root of 2.) Let $\mathcal{U}$ denote the union of the set of negative rational numbers and the set of all rational numbers x such that x^2 is less than 2. Prove that $\mathcal{U}$ is a real number and that the product $\mathcal{U}\mathcal{U}$ is the real number corresponding to the rational number 2.

A very nice and complete exposition of the Dedekind cuts construction of the real numbers can be found in "Calculus" by Michael Spivak (Publish or Perish, Inc., Houston, Texas), which also contains a beautiful treatment of the principles of calculus.

Chapter 9
The Complex Numbers

The set of real numbers is rich enough to be useful in a wide variety of situations. In particular, it provides a number for every distance. There are, however, some situations where additional numbers are required.

9.1 What is a Complex Number?

Let's consider the problem of finding roots for polynomial equations. Recall that polynomials are expressions such as $7x^2 + 5x - 3$, and $\sqrt{2}x^3 + \frac{5}{7}x$, and $x^7 - 1$. The general definition is the following.

Definition 9.1.1. A *polynomial* is an expression of the form

$$a_n x^n + a_{n-1} x^{n-1} + \cdots + a_1 x + a_0$$

where n is a natural number and the a_i are numbers with $a_n \neq 0$, or simply a number a_0 (which we call a *constant polynomial*). The a_i are called the *coefficients* of the polynomial. The natural number n, the highest power to which x occurs in the polynomial, is called the *degree* of the polynomial. A constant polynomial is said to have degree 0.

Note that in the definition of polynomial we used x as the variable; this is very standard. However, it is often the case that other variables are used as well. For example, $z^3 - 4z + 3$ would be a polynomial in the variable z.

A polynomial defines a function; whenever any specific number is substituted for x, the resulting expression is a number. The values of x for which the polynomial is 0 have special significance.

Definition 9.1.2. A *root* or *zero* of the polynomial $a_n x^n + a_{n-1} x^{n-1} + \cdots + a_1 x + a_0$ is a number that when substituted for x makes

$$a_n x^n + a_{n-1} x^{n-1} + \cdots + a_1 x + a_0 = 0$$

D. Rosenthal et al., *A Readable Introduction to Real Mathematics*,
Undergraduate Texts in Mathematics, DOI 10.1007/978-3-319-05654-8_9,
© Springer International Publishing Switzerland 2014

For example, 2 is a root of the polynomial $x^2 - 4$, 3 is a root of the polynomial $5x^2 - 2x - 39$, $-\frac{7}{5}$ is a root of the polynomial $5x + 7$, and so on.

A very natural question is: Which polynomials have roots? All polynomials of degree 1 have roots: the polynomial $a_1 x + a_0$ has the root $-\frac{a_0}{a_1}$. What about polynomials of degree 2? A simple example is the polynomial $x^2 + 1$. No real number is a root of that polynomial, since x^2 is nonnegative for every real number x, and therefore $x^2 + 1$ is strictly greater than 0 for every real x. If the polynomial $x^2 + 1$ is to have a root, it would have to be in a larger number system than that of the real numbers. Such a system was invented by mathematicians hundreds of years ago.

We use the symbol i to denote a root of the polynomial $x^2 + 1$. That is, we define i^2 to be equal to -1. We then combine this symbol i with real numbers, using standard manipulations of algebra in the usual ways, to get the complex numbers. The definition is the following.

Definition 9.1.3. A *complex number* is an expression of the form $a + bi$ where a and b are real numbers. The real number a is called the *real part* of $a + bi$ and the real number b is called the *imaginary part* of $a + bi$. We sometimes use the notation $\text{Re}(z)$ and $\text{Im}(z)$ to denote the real and imaginary parts of the complex number z, respectively. Addition of complex numbers is defined by

$$(a + bi) + (c + di) = (a + c) + (b + d)i$$

Multiplication of complex numbers is defined by

$$(a + bi)(c + di) = ac + adi + bic + bdi^2 = ac + bdi^2 + (ad + bc)i$$
$$= (ac - bd) + (ad + bc)i$$

where we replaced i^2 by -1 to get the last equation.

Example 9.1.4.

$$(6 + 2i) + (-4 + 5i) = 2 + 7i$$

$$(-\sqrt{12} + \sqrt{6}i) + (4 + \pi i) = (-\sqrt{12} + 4) + (\sqrt{6} + \pi)i$$

$$(7+2i)(3-4i) = 21+6i-28i-8i^2 = 21-22i-8(-1) = 21+8-22i = 29-22i$$

Notation 9.1.5. The set of all complex numbers is denoted by $\mathbb{C}$.

We use the symbol 0 as an abbreviation for the complex number $0 + 0i$. More generally, we use a as an abbreviation for the complex number $a + 0i$. Similarly we use bi as an abbreviation for the complex number $0 + bi$. When r is a real number, then $r(a + bi)$ is simply $ra + rbi$.

Note that every complex number has an additive inverse (i.e., a complex number that gives 0 when added to the given number). For example, the additive inverse of $-7 + \sqrt{2}i$ is $7 - \sqrt{2}i$. In general, the additive inverse of $a + bi$ is $-a + (-b)i$.

Definition 9.1.6. The number $a - bi$ is called the *complex conjugate* of the number $a + bi$; the complex conjugate of a complex number is often denoted by placing a horizontal bar over the complex number:

$$\overline{a + bi} = a - bi$$

Example 9.1.7. The complex conjugate of $2 + 3i$ is $2 - 3i$, or $\overline{2 + 3i} = 2 - 3i$. Similarly, $\overline{-\sqrt{3} - 5i} = -\sqrt{3} + 5i$, and $\overline{9} = 9$.

The product of a complex number and its conjugate is important.

Theorem 9.1.8. *For any complex number $a + bi$, $(a + bi)(a - bi) = a^2 + b^2$.*

Proof. Simply multiplying gives the result; we invite you to verify this in Problem 5 at the end of this chapter.   □

Definition 9.1.9. The *modulus* of the complex number $a + bi$ is $\sqrt{a^2 + b^2}$; it is often denoted $|a + bi|$.

Thus, $(a + bi)(\overline{a + bi}) = |a + bi|^2$.

Do complex numbers have multiplicative inverses? That is, given $a + bi$, is there a complex number $c + di$ such that $(a + bi)(c + di) = 1$? Of course, the complex number 0 cannot have a multiplicative inverse, since its product with any complex number is 0. What about other complex numbers?

Given a complex number $a + bi$, let's try to compute a multiplicative inverse $c + di$ for it. Suppose that $(a + bi)(c + di) = 1$. Multiplying both sides of this equation by $\overline{a + bi}$ and using the fact that $(a + bi)(\overline{a + bi}) = a^2 + b^2$ yields $(a^2 + b^2)(c + di) = a - bi$. Since $a^2 + b^2$ is a real number, this implies (unless $a^2 + b^2 = 0$) that $c + di = \frac{a}{a^2+b^2} - \frac{b}{a^2+b^2}i$. Note that if $a^2 + b^2 = 0$, then $a = b = 0$, so the number $a + bi$ is 0. Thus, if $a + bi$ has a multiplicative inverse, that multiplicative inverse must be $\frac{a}{a^2+b^2} - \frac{b}{a^2+b^2}i$. In fact, as we now show, that expression is a multiplicative inverse for $a + bi$.

Theorem 9.1.10. *If $a + bi \neq 0$, then $\frac{a}{a^2+b^2} - \frac{b}{a^2+b^2}i$ is a multiplicative inverse for $a + bi$.*

Proof. We verify this by simply multiplying

$$(a + bi) \cdot \left(\frac{a}{a^2 + b^2} - \frac{b}{a^2 + b^2}i \right) = \frac{a^2}{a^2 + b^2} + \frac{b^2}{a^2 + b^2} - \frac{ab}{a^2 + b^2}i + \frac{ab}{a^2 + b^2}i$$

which simplifies to $\frac{a^2+b^2}{a^2+b^2} = 1$.   □

As with real numbers, the multiplicative inverse of the complex number $a + bi$ is often denoted $\frac{1}{a+bi}$.

9.2   The Complex Plane

It is very useful to represent complex numbers in a coordinatized plane. We let the complex number $a + bi$ correspond to the point (a, b) in the ordinary xy-plane. Note that the modulus $|a + bi|$ is the distance from (a, b) to the origin.

Definition 9.2.1. For a complex number $a + bi$ other than 0, the angle that the line from $(0, 0)$ to (a, b) makes (in a counterclockwise direction) with the positive x-axis is called the *argument* of $a + bi$.

In day to day life, angles are usually measured in degrees: a right angle is 90°, a straight angle is 180°, and an angle of 37° is $\frac{37}{180}$ of a straight angle. For doing mathematics, however, it is almost always more convenient to measure angles differently.

Definition 9.2.2. The *radian measure* of the angle θ is the length of the arc of a circle of radius 1 that is cut off by an angle θ at the center of the circle. (See Figure 9.1 below.)

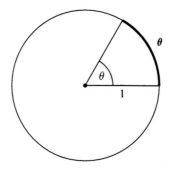

Fig. 9.1   The radian measure of an angle

Thus, since a circle of radius 1 has circumference 2π, the radian measure of a right angle is $\frac{\pi}{2}$, of a straight angle is π, of an angle of 60° is $\frac{\pi}{3}$, and so on. Note that 2π is a full revolution. Therefore, for any natural number k, the angle $\theta + 2\pi k$ measured from the positive x-axis ends up at the same position as θ. We will use the radian measure of angles for the rest of this chapter.

We require the basic properties of the trigonometric functions sine, cosine, and tangent.

If the complex number $a + bi$ has modulus r and argument θ, then $a = r \cos \theta$, and $b = r \sin \theta$. To see this, first consider the case where both a and b are greater than or equal to 0, which is equivalent to θ being an angle between 0 and $\frac{\pi}{2}$. Then the situation is as in Figure 9.2 below. The fact that the cosine of an angle in a right triangle is its adjacent side divided by its hypotenuse gives $\cos \theta = \frac{a}{r}$, or $a = r \cos \theta$. Similarly, the fact that the sine of θ is the opposite side divided by

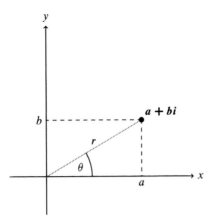

Fig. 9.2 Representations of a complex number

the hypotenuse gives $\sin \theta = \frac{b}{r}$, or $b = r \sin \theta$. Similar analysis yields the same equations when one or more of a and b is negative, and thus the conclusion holds for any θ. Each complex number is determined by its modulus r and its argument θ; that is, $a + bi = r(\cos \theta + i \sin \theta)$. (The only complex number whose modulus is 0 is the number 0, and 0 is the only complex number whose argument is not defined.)

Definition 9.2.3. The *polar form* of the complex number with modulus r and argument θ is $r(\cos \theta + i \sin \theta)$.

One reason that the polar form is important is because there is a neat description of multiplication of complex numbers in terms of their moduli and arguments.

Theorem 9.2.4. *The modulus of the product of two complex numbers is the product of their moduli. The argument of the product of two complex numbers is the sum of their arguments.*

Proof. Simply multiplying the two complex numbers $r_1(\cos \theta_1 + i \sin \theta_1)$ and $r_2(\cos \theta_2 + i \sin \theta_2)$ and collecting terms yields

$$r_1 r_2 \big((\cos \theta_1 \cos \theta_2 - \sin \theta_1 \sin \theta_2) + i(\cos \theta_1 \sin \theta_2 + \sin \theta_1 \cos \theta_2)\big)$$

Recall the addition formulae for cosine and sine:

$$\cos(\theta_1 + \theta_2) = \cos \theta_1 \cos \theta_2 - \sin \theta_1 \sin \theta_2$$

and

$$\sin(\theta_1 + \theta_2) = \sin\theta_1 \cos\theta_2 + \sin\theta_2 \cos\theta_1$$

Using these addition formulae in the above equation shows that the product is equal to

$$r_1 r_2\big(\cos(\theta_1 + \theta_2) + i\sin(\theta_1 + \theta_2)\big)$$

This proves the theorem.   □

Thus, to multiply two complex numbers, we can simply multiply their moduli and add their arguments. In particular, the case where the two complex numbers are equal shows that the square of a complex number is obtained by squaring its modulus and doubling its argument. One application of this fact is the following.

Theorem 9.2.5. *Every complex number has a complex square root.*

Proof. To show that any given complex number has a square root, write the number in polar form, say $z = r(\cos\theta + i\sin\theta)$. Let w equal $\sqrt{r}(\cos\frac{\theta}{2} + i\sin\frac{\theta}{2})$. By the previous theorem, $w^2 = z$.   □

It is also easy to compute powers higher than 2.

De Moivre's Theorem 9.2.6. *For every natural number n*

$$\big(r(\cos\theta + i\sin\theta)\big)^n = r^n(\cos n\theta + i\sin n\theta)$$

Proof. This is easily established by induction on n. The case $n = 1$ is clear. Suppose that the formula holds for $n = k$, that is, suppose

$$\big(r(\cos\theta + i\sin\theta)\big)^k = r^k(\cos k\theta + i\sin k\theta)$$

Multiplying both sides of this equation by $r(\cos\theta + i\sin\theta)$ and using Theorem 9.2.4 gives

$$\big(r(\cos\theta + i\sin\theta)\big)^{k+1} = r^k(\cos k\theta + i\sin k\theta)\cdot r(\cos\theta + i\sin\theta)$$
$$= r\cdot r^k(\cos(k\theta + \theta) + i\sin(k\theta + \theta))$$
$$= r^{k+1}\big(\cos((k+1)\theta) + i\sin((k+1)\theta)\big)$$

This is the formula for $n = k + 1$, so the theorem is established by mathematical induction.   □

De Moivre's Theorem leads to some very nice computations, such as the following.

Example 9.2.7. We can compute $(1+i)^8$ as follows. First, $|1+i| = \sqrt{2}$. Plotting $1+i$ as the point $(1, 1)$ in the plane makes it apparent that the argument of $1+i$ is $\frac{\pi}{4}$. Thus, by De Moivre's Theorem (9.2.6), the modulus of $(1+i)^8$ is $(\sqrt{2})^8 = 2^4 = 16$ and the argument is $8 \cdot \frac{\pi}{4} = 2\pi$. It follows that

$$(1+i)^8 = 16(\cos 2\pi + i \sin 2\pi) = 16$$

Therefore, $(1+i)^8 = 16$.

The following is a very similar computation.

Example 9.2.8.

$$(1+i)^{100} = \left(\sqrt{2}\left(\cos\frac{\pi}{4} + i \sin\frac{\pi}{4}\right)\right)^{100} = 2^{50}(\cos 25\pi + i \sin 25\pi)$$

Since the angle with the positive x-axis of 25π radians is in the same position as the angle of π radians, it follows that

$$(1+i)^{100} = 2^{50}(\cos \pi + i \sin \pi) = 2^{50}(-1+0) = -2^{50}$$

It is interesting to compute the roots of the complex number 1. The number 1 is sometimes called *unity* in this context.

Example 9.2.9 (Square Roots of Unity). Obviously, $1^2 = 1$ and $(-1)^2 = 1$. Are there any other complex square roots of 1? To compute the square roots of 1 we can proceed as follows. Let $z = r(\cos\theta + i\sin\theta)$. Then $z^2 = r^2(\cos 2\theta + i\sin 2\theta)$. If $z^2 = 1$, then r^2 must be the modulus of 1; i.e., $r^2 = 1$. Since r is nonnegative, it follows that $r = 1$. Also, $\cos 2\theta + i \sin 2\theta = 1$. Therefore, $\cos 2\theta = 1$ and $\sin 2\theta = 0$. What are the possible values of θ? Clearly, $\theta = 0$ is one solution, as is $\theta = \pi$; the corresponding values of z are $z = \cos 0 + i \sin 0 = 1$ and $z = \cos \pi + i \sin \pi = -1$. Are there any other possible values of θ? Of course there are: θ could be 2π or 3π or 4π or 5π. If θ is any multiple of π, then $\cos 2\theta = 1$ and $\sin 2\theta = 0$. However, we do not get any new values of z by using those other values of θ. We only get $z = 1$ or $z = -1$ depending upon whether we have an even or an odd multiple of π. It is easily seen that only the multiples of π simultaneously satisfy the equations $\cos 2\theta = 1$ and $\sin 2\theta = 0$. This follows from the fact that $\cos\phi = 1$ only when ϕ is a multiple of 2π, so $\cos 2\theta = 1$ only when θ is a multiple of π. Thus, the only complex square roots of 1 are 1 and -1.

Cube roots of unity are more interesting. The only real number z satisfying $z^3 = 1$ is $z = 1$. However, there are other complex numbers satisfying this equation.

Example 9.2.10 (Cube Roots of Unity). Suppose that $z = r(\cos\theta + i\sin\theta)$ and $z^3 = 1$. Then clearly $r = 1$. By De Moivre's Theorem (9.2.6), $z^3 = \cos 3\theta + i \sin 3\theta$. From $z^3 = 1$ we get $\cos 3\theta = 1$ and $\sin 3\theta = 0$. These equations are, of course, satisfied by $\theta = 0$, which gives $z = \cos 0 + i \sin 0 = 1$, the obvious cube root of 1. But also $\cos 3\theta = 1$ and $\sin 3\theta = 0$ when $3\theta = 2\pi$. That is, when $\theta = \frac{2\pi}{3}$. Thus, $z = \cos \frac{2\pi}{3} + i \sin \frac{2\pi}{3} = -\frac{1}{2} + \frac{\sqrt{3}}{2}i$ is another cube root of 1. (Once it is conjectured that $-\frac{1}{2} + \frac{\sqrt{3}}{2}i$ is a cube root of 1, that could be verified by simply computing $\left(-\frac{1}{2} + \frac{\sqrt{3}}{2}i\right)^3$.) There is another cube root of 1. If $3\theta = 4\pi$, then $\cos 3\theta = 1$ and $\sin 3\theta = 0$. Thus, $z = \cos \frac{4\pi}{3} + i \sin \frac{4\pi}{3} = -\frac{1}{2} - \frac{\sqrt{3}}{2}i$ is another cube root of 1. Therefore, we have found three cube roots of 1: 1, $-\frac{1}{2} + \frac{\sqrt{3}}{2}i$ and $-\frac{1}{2} - \frac{\sqrt{3}}{2}i$. Are there any other cube roots of 1? If $3\theta = 6\pi$, then $\cos 3\theta = 1$ and $\sin 3\theta = 0$. When $3\theta = 6\pi$, $\theta = 2\pi$. Thus, $\cos \theta + i \sin \theta$ is simply 1, so we are not getting an additional cube root. More generally, for every integer k, $\cos 2k\pi = 1$ and $\sin 2k\pi = 0$. However, if $3\theta = 2k\pi$, then there are only the three different values given above for $\cos \theta + i \sin \theta$, since all the values of θ obtained from other values of k differ from one of 0, $\frac{4\pi}{3}$ and $\frac{2\pi}{3}$ by a multiple of 2π.

It is interesting to plot the three cube roots of unity in the plane.

The three cube roots of unity are obtained by starting at the point 1 on the circle of radius 1 and then moving in a counterclockwise direction $\frac{2\pi}{3}$ to get the next cube root and then moving an additional $\frac{2\pi}{3}$ to get the third cube root (Figure 9.3).

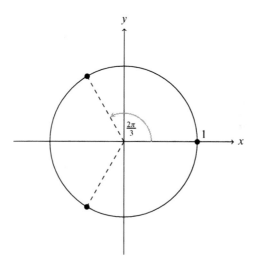

Fig. 9.3   The cube roots of 1

For each natural number n, the complex n^{th} roots of 1 can be obtained by starting at 1 and successively moving around the unit circle in a counterclockwise direction through angles of $\frac{2\pi}{n}$.

Example 9.2.11 (n^{th} Roots of Unity).   For each natural number n, the complex n^{th} roots of 1 are the numbers 1, $\cos \frac{2\pi}{n} + i \sin \frac{2\pi}{n}$, $\cos \frac{4\pi}{n} + i \sin \frac{4\pi}{n}$, $\cos \frac{6\pi}{n} + i \sin \frac{6\pi}{n}$, $\cos \frac{8\pi}{n} + i \sin \frac{8\pi}{n}, \ldots, \cos \frac{2\pi(n-1)}{n} + i \sin \frac{2\pi(n-1)}{n}$.

To see this, first note that, for any natural number k,

$$\left(\cos\frac{2\pi k}{n} + i\sin\frac{2\pi k}{n}\right)^n = \cos 2\pi k + i\sin 2\pi k$$

by De Moivre's Theorem (9.2.6). Since $\cos 2\pi k + i\sin 2\pi k = 1$, this shows that each of $\cos\frac{2\pi k}{n} + i\sin\frac{2\pi k}{n}$ is an nth root of unity. To show that these are the only n^{th} roots of unity we proceed as follows. Suppose that $z = \cos\theta + i\sin\theta$ and $z^n = 1$. Then $\cos n\theta + i\sin n\theta = 1$, so $\cos n\theta = 1$ and $\sin n\theta = 0$. Thus, $n\theta = 2\pi k$ for some integer k. It follows that $\theta = \frac{2\pi k}{n}$. Taking $k = 0, 1, \ldots, n-1$ gives the nth roots that we have listed. Taking other values of k gives different values for $\frac{2\pi k}{n}$, but each of them differs from one of the listed values by a multiple of 2π and therefore gives a value for $\cos\theta + i\sin\theta$ that we already have. Thus, the n roots that we listed are all of the nth roots of unity.

Roots of other complex numbers can also be computed.

Example 9.2.12. All of the solutions of the equation $z^3 = 1 + i$ can be found as follows. First note that $|1 + i| = \sqrt{2}$ and the argument of $1 + i$ is $\frac{\pi}{4}$. That is, $1 + i = \sqrt{2}\left(\cos\frac{\pi}{4} + i\sin\frac{\pi}{4}\right)$. Suppose that $z = r(\cos\theta + i\sin\theta)$ and $z^3 = 1 + i$. Then $z^3 = r^3(\cos 3\theta + i\sin 3\theta)$. Therefore $r^3 = \sqrt{2}$, so $r = 2^{\frac{1}{6}}$, and 3θ is $\frac{\pi}{4}$ or $\frac{\pi}{4} + 2\pi$ or $\frac{\pi}{4} + 4\pi$. Therefore, θ itself can be $\frac{\pi}{12}$, $\frac{3\pi}{4}$, or $\frac{17\pi}{12}$. This gives the three solutions of the equation $z^3 = 1 + i$: $2^{\frac{1}{6}}\left(\cos\frac{\pi}{12} + i\sin\frac{\pi}{12}\right)$, $2^{\frac{1}{6}}\left(\cos\frac{3\pi}{4} + i\sin\frac{3\pi}{4}\right)$, and $2^{\frac{1}{6}}\left(\cos\frac{17\pi}{12} + i\sin\frac{17\pi}{12}\right)$.

9.3 The Fundamental Theorem of Algebra

One reason for introducing complex numbers was to provide a root for the polynomial $x^2 + 1$. There are many other polynomials that do not have any real roots. For example, if $p(x)$ is any polynomial, then the polynomial obtained by writing out $(p(x))^2 + 1$ has no real roots, since its value is at least 1 for every value of x.

Does every such polynomial have a complex root? More generally, does every polynomial have a complex root? There is a trivial sense in which the answer to this question is "no," since constant polynomials other than 0 clearly do not have any roots of any kind. For other polynomials, the answer is not so simple. It is a remarkable and very useful fact that every non-constant polynomial with real coefficients, or even with complex coefficients, has a complex root.

The Fundamental Theorem of Algebra 9.3.1. *Every non-constant polynomial with complex coefficients has a complex root.*

There are a number of different proofs of the Fundamental Theorem of Algebra. They all rely on mathematical concepts that we do not develop in this book. We will therefore simply discuss implications of this theorem without proving it.

How many roots does a polynomial have?

Example 9.3.2. The only root of the polynomial $p(z) = z^2 - 6z + 9$ is $z = 3$. This follows from the fact that $p(z) = (z - 3)(z - 3)$. Since the product of two complex numbers is 0 only if at least one of the numbers is 0, the only solution to $p(z) = 0$ is $z = 3$. In some sense, however, this polynomial has 3 as a "double root"; we'll discuss this a little more below.

To explore the question of the number of roots that a polynomial can have, we need to use the concept of the division of one polynomial by another. This concept of division is very similar to "long division" of one natural number into another. Actually, we only need a special case of this concept, the case where the polynomial divisor is linear (i.e., has degree 1). We begin with an example.

Example 9.3.3. To divide $z - 3$ into $z^4 + 5z^3 - 2z + 1$ we proceed as follows:

$$
\begin{array}{r}
z^3 \;+8z^2 \;+24z \;+70 \\
z - 3 \,\overline{) \,z^4 \;+5z^3 \qquad\quad -2z \quad +1} \\
\underline{z^4 \;-3z^3} \\
8z^3 \qquad\qquad -2z \quad +1 \\
\underline{8z^3 \;-24z^2} \\
24z^2 \;-2z \quad +1 \\
\underline{24z^2 \;-72z} \\
70z \quad +1 \\
\underline{70z \;-210} \\
211
\end{array}
$$

What this tabulation shows, like with long division of numbers, is that

$$z^4 + 5z^3 - 2z + 1 = (z - 3)(z^3 + 8z^2 + 24z + 70) + 211$$

The only consequence of the division of one polynomial by another that we need for present purposes is the following.

Theorem 9.3.4. *If r is a complex number and p(z) is a non-constant polynomial with complex coefficients, then there exists a polynomial q(z) and a constant c such that*

$$p(z) = (z - r)q(z) + c$$

Proof. We will proceed by using the Principle of Complete Mathematical Induction
(2.2.1) on the degree of the polynomial $p(z)$. Since $p(z)$ is non-constant, the base
case of our induction proof is when the degree of $p(z)$ is 1. In other words, $p(z) =
az + b$, where a and b are complex numbers and $a \neq 0$. Let r be a complex number.
As in Example 9.3.3, we will use long division to divide $z - r$ into $p(z)$:

$$
\begin{array}{r}
a \\
z - r \overline{\smash{)}\, az \quad +b} \\
\underline{az \quad -ar} \\
ar + b
\end{array}
$$

This shows that $p(z) = az + b = (z - r) \cdot a + (ar + b)$. So setting $q(z) = a$ and
$c = ar + b$ gives us the desired result when the degree of $p(z)$ is 1. Thus, the base
case of the induction is complete.

Now assume that the theorem is true for all polynomials of degree less than or
equal to $n \geq 1$. Using this assumption we will show that the theorem holds for
any polynomial of degree equal to $n + 1$. Let $p(z)$ be a polynomial with complex
coefficients with degree $n + 1$. That is,

$$p(z) = a_{n+1}z^{n+1} + a_n z^n + a_{n-1}z^{n-1} + \cdots + a_1 z + a_0$$

where each a_i is a complex number and a_{n+1} is nonzero. Let r be a complex number.
Once again we use long division to divide $z - r$ into $p(z)$:

$$
\begin{array}{r}
a_{n+1}z^n \\
z - r \overline{\smash{)}\, a_{n+1}z^{n+1} \qquad +a_n z^n +a_{n-1}z^{n-1} + \cdots + a_1 z + a_0} \\
\underline{a_{n+1}z^{n+1} \qquad -ra_{n+1}z^n} \\
(a_n + ra_{n+1})z^n +a_{n-1}z^{n-1} + \cdots + a_1 z + a_0
\end{array}
$$

To simplify the notation, let $p_n(z) = (a_n + ra_{n+1})z^n + a_{n-1}z^{n-1} + \cdots + a_1 z + a_0$. Then
the above long division tells us that $p(z) = (z - r)(a_{n+1}z^n) + p_n(z)$. Since $p_n(z)$ is
a polynomial of degree less than or equal to n, the induction hypothesis tells us that
there exists a polynomial $q_n(z)$ and a constant c such that $p_n(z) = (z - r)q_n(z) + c$.
Thus,

$$
\begin{aligned}
p(z) &= (z - r)(a_{n+1}z^n) + p_n(z) &= (z - r)(a_{n+1}z^n) + (z - r)q_n(z) + c \\
&= (z - r)\big(a_{n+1}z^n + q_n(z)\big) + c
\end{aligned}
$$

Therefore, setting $q(z) = a_{n+1}z^n + q_n(z)$ gives us the desired result when the degree
of $p(z)$ is $n + 1$. This completes the proof by induction. $\qquad\square$

Definition 9.3.5. The polynomial $f(z)$ is a *factor* of the polynomial $p(z)$ if there exists a polynomial $q(z)$ such that $p(z) = f(z)q(z)$.

The Factor Theorem 9.3.6. *The complex number r is a root of a polynomial $p(z)$ if and only if $z - r$ is a factor of $p(z)$.*

Proof. If $(z - r)$ is a factor of $p(z)$, then $p(z) = (z - r)q(z)$ implies that $p(r) = (r - r)q(r) = 0 \cdot q(r) = 0$. Conversely, suppose that r is a root of $p(z)$. By Theorem 9.3.4, $p(z) = (z - r)q(z) + c$ for some constant c. Substituting r for z and using the fact that r is a root gives $0 = (r - r)q(r) + c$, so $0 = 0 + c$, from which it follows that $c = 0$. Hence, $p(z) = (z - r)q(z)$ and $z - r$ is a factor of $p(z)$.   □

Example 9.3.7. The complex number $2i$ is a root of the polynomial $iz^3 + z^2 - 4$ (as can be seen by simply substituting $2i$ for z in the expression for the polynomial and noting that the result is 0). It follows from the Factor Theorem that $z - 2i$ is a factor of the given polynomial. Doing "long division" gives $iz^3 + z^2 - 4 = (z - 2i)(iz^2 - z - 2i)$.

We can use the Factor Theorem to determine the maximum number of roots that a polynomial may have.

Theorem 9.3.8. *A polynomial of degree n has at most n complex roots; if "multiplicities" are counted, it has exactly n roots.*

Proof. Let $p(z)$ be a polynomial of degree n. If n is at least 1, then $p(z)$ has a root, say r_1, by the Fundamental Theorem of Algebra (9.3.1). By the Factor Theorem (9.3.6), there exists a polynomial $q_1(z)$ such that $p(z) = (z - r_1)q_1(z)$. The degree of q_1 is clearly $n - 1$. If $n - 1 > 0$, then $q_1(z)$ has a root, say r_2. It follows from the Factor Theorem that there is a polynomial $q_2(z)$ such that $q_1(z) = (z - r_2)q_2(z)$. The degree of $q_2(z)$ is $n - 2$, and

$$p(z) = (z - r_1)(z - r_2)q_2(z)$$

This process can continue (a formal proof can be given using mathematical induction) until a quotient is simply a constant, say k. Then,

$$p(z) = k(z - r_1)(z - r_2) \cdots (z - r_n)$$

If the r_i are all different, the polynomial will have n roots. If some of the r_i coincide, collecting all the terms where r_i is equal to a given r produces a factor of the form $(z - r)^m$, where m is the number of times that r occurs in the factorization. In this situation, we say that r is a *root of multiplicity m* of the polynomial. Thus, a polynomial of degree n has at most n distinct roots. If the roots are counted according to their multiplicities, then a polynomial of degree n has exactly n roots.   □

9.4 Problems

Basic Exercises

1. Write the following complex numbers in $a + bi$ form, where a and b are real numbers:

   (a) $\left(\frac{1}{\sqrt{2}} + \frac{i}{\sqrt{2}}\right)^{10}$

   (b) $\left(\frac{1}{\sqrt{2}} + \frac{i}{\sqrt{2}}\right)^{106}$

   (c) $\left(-\frac{\sqrt{3}}{2} + \frac{i}{2}\right)^{11}$

   (d) $\frac{3+2i}{-1-i}$

   (e) $\frac{3}{\sqrt{2}+i}$

   (f) i^{574}

   (g) i^{575}

   (h) $\frac{1}{i^9}$

2. Show that the real part of $(1 + i)^{10}$ is 0.
3. Find both square roots of the following numbers:

   (a) $-i$

   (b) $-15 - 8i$

4. Find the cube roots of the following numbers:

   (a) 2

   (b) $8\sqrt{3} + 8i$

5. Prove Theorem 9.1.8; that is, prove that for any complex number $a + bi$,

$$(a + bi)(a - bi) = a^2 + b^2$$

Interesting Problems

6. Prove the *Quadratic Formula*; i.e., prove that the polynomial $az^2 + bz + c$, where a, b and c are any complex numbers and a is different from 0, has roots $z = \frac{-b \pm \sqrt{b^2 - 4ac}}{2a}$.

   [Hint: Rewrite the equation as $z^2 + \frac{b}{a}z + \frac{c}{a} = 0$, and use the fact that $\left(z + \frac{b}{2a}\right)^2 = z^2 + \frac{b}{a}z + \frac{b^2}{4a^2}$.]

7. Find all solutions to the equation $iz^2 + 2z + i = 0$.
8. Find a polynomial p with integer coefficients such that $p\left(3 + i\sqrt{7}\right) = 0$.
9. Find all the complex roots of the polynomial $z^6 + z^3 + 1$.
10. Find all the complex roots of the polynomial $z^7 - z$.
11. Find a polynomial whose complex roots are $2 - i, 2 + i, 7$.

Challenging Problems

12. Find all the complex solutions of $\dfrac{z^3 + 1}{z^3 - 1} = i$.

13. Let p be a polynomial with real coefficients. Prove that the complex conjugate of each root of p is also a root of p.

    [Hint: First show that for any two complex numbers the sum of the conjugates is the conjugate of the sum, and the product of the conjugates is the conjugate of the product.]

14. Show that every non-constant polynomial with real coefficients can be factored into a product of linear (i.e., of degree 1) and quadratic (i.e., of degree 2) polynomials, each of which also has real coefficients.

15. Extend De Moivre's Theorem (9.2.6) to prove that

$$\left(r(\cos\theta + i\sin\theta)\right)^n = r^n(\cos n\theta + i\sin n\theta)$$

    for negative integers n.

Chapter 10
Sizes of Infinite Sets

How many natural numbers are there? How many even natural numbers are there? How many odd natural numbers are there? How many rational numbers are there? How many real numbers are there? How many points are there in the plane? How many sets of natural numbers are there? How many different circles are there in the plane? One answer to all the above questions would be: there are an infinite number of them. But there are more precise answers that can be given; there are, in a sense that we will explain, an infinite number of different size infinities.

10.1 Cardinality

Definition 10.1.1. By a *set* we simply mean any collection of things; the things are called *elements of the set*. (As will be discussed at the end of this chapter, such a general definition of set is problematic in certain senses.)

For example, the collection of all words on this page is a set. The collection containing the letters a, b, and c is a set: it could be denoted $S = \{a, b, c\}$. The set of all real numbers greater than 4 could be written:

$$\{x : x > 4\}$$

The fact that something is an element of a set is often denoted with the Greek letter epsilon, $\in$. We write $x \in S$ to represent the fact that x is an element of the set S.

Definition 10.1.2. If S is a set, a *subset* of S is a set all of whose elements are elements of the set S. The notation $T \subset S$ is used to signify that T is a subset of S. The *empty set* is the set that has no elements at all. It is denoted $\emptyset$. The empty set is, by definition, a subset of every set. That is, $\emptyset \subset S$ for every set S. The *union* of a collection of sets is the set consisting of all elements that occur in any of the given sets. The union of sets S and T is denoted $S \cup T$ and similar notation is used for the union of more than two sets. The *intersection* of a collection of sets

D. Rosenthal et al., *A Readable Introduction to Real Mathematics*,
Undergraduate Texts in Mathematics, DOI 10.1007/978-3-319-05654-8_10,
© Springer International Publishing Switzerland 2014

is the set consisting of all elements that are in every set in the given collection. The intersection of the sets S and T is denoted $S \cap T$ and similar notation is used for the intersection of more than two sets. If the intersection of two sets is the empty set, the sets are said to be *disjoint*.

How should we define the concept that two sets have the same number of elements? For finite sets, we count the number of elements in each set. When we count the number of elements in a set, we assign the number 1 to one of the elements of the set, then assign the number 2 to another element of the set, then 3 to another element of the set, and so on, until we have counted every element in the set. If the set has n elements, when we finish counting we will have assigned a number in the set $\{1, 2, 3, \ldots, n\}$ to each element of the set and will not have assigned two different numbers to the same element in the set. That is, counting that a set has n elements produces a pairing of the elements of the set $\{1, 2, 3, \ldots, n\}$ with the elements of the set that we are counting. A set whose elements can be paired with the elements of the set $\{1, 2, 3, \ldots, n\}$ is said to have n elements.

More generally, we can say that two sets have the same number of elements if the elements of those two sets can be paired with each other.

Example 10.1.3. Pairs of running shoes are manufactured in a given factory. Each day, some number of pairs is manufactured. Even without knowing how many pairs were manufactured in a given day, we can still conclude that the same number of left shoes was manufactured as the number of right shoes that was manufactured, since they are manufactured in pairs. If, for example, the number of left shoes was determined to be 1012, then it could be concluded that the number of right shoes was also 1012. This could be established as follows: since the set $\{1, 2, 3, \ldots, 1012\}$ can be paired with the set of left shoes, it could also be paired with the set of right shoes, simply by pairing each right shoe to the number assigned to the corresponding left shoe in the pair.

The above discussion suggests the general definition that we shall use. In the following, the phrase "have the same cardinality" is the standard mathematical terminology for what might colloquially be expressed "have the same size."

Definition 10.1.4 (Rough Definition). The sets S and T are said to *have the same cardinality* if their elements can be paired with each other's.

We need to be able to precisely define what we mean by a "pairing" of the elements of two sets. This can be specified in terms of functions. A *function* from a set S into a set T is simply an assignment of one element of T to each element of S. For example, if $S = \{a, b, d, e\}$ and $T = \{+, \pi\}$, then one particular function taking S to T is the function f defined by $f(a) = \pi$, $f(b) = \pi$, $f(d) = +$, and $f(e) = \pi$.

Definition 10.1.5. The notation $f : S \rightarrow T$ is used to denote a function f taking the set S into the set T; that is, a mapping of each element of S to an element of T. The set S is called the *domain* of the function. The *range* of a function is the set of all its values; that is, the range of $f : S \rightarrow T$ is $\{f(s) : s \in S\}$.

Definition 10.1.6. A function $f : S \to T$ is *one-to-one* (or *injective*) if $f(s_1) \neq f(s_2)$ whenever $s_1 \neq s_2$. That is, a function is one-to-one if it does not send two different elements to the same element.

We also require another property that functions may have.

Definition 10.1.7. A function $f : S \to T$ is *onto* (or *surjective*) if for every $t \in T$ there is an $s \in S$ such that $f(s) = t$; that is, the range of f is all of T.

Note that a one-to-one, onto function from a set S onto a set T gives a pairing of the elements of S with those of T.

The formal definition of when sets are to be considered to have the same size can be stated as follows.

Definition 10.1.8. The sets S and T *have the same cardinality* if there is a function $f : S \to T$ that is one-to-one and onto all of T.

We require the concept of the inverse of a function. If f is a one-to-one function mapping a set S onto a set T, then there is a function mapping T onto S that "sends elements back to where they came from," via f.

Definition 10.1.9. If f is a one-to-one function mapping S onto T, then the *inverse of f*, often denoted f^{-1}, is the function mapping T onto S defined by $f^{-1}(t) = s$ when $f(s) = t$.

With respect to the definition above, note that f must be onto for f^{-1} to be defined on all of T. Also, f must be one-to-one; otherwise for some t there will be more than one s for which $f(s) = t$ and therefore $f^{-1}(t)$ will not be determined. If f is a one-to-one function mapping S onto T, then f^{-1} is a one-to-one function mapping T onto S.

Let's consider some examples.

Example 10.1.10. The set of even natural numbers and the set of odd natural numbers have the same cardinality.

Proof. Write the set of even natural numbers as $\mathcal{E} = \{2, 4, \ldots, 2n, \ldots\}$ and the set of odd natural numbers as $\mathcal{O} = \{1, 3, \ldots, 2n + 1, \ldots\}$. We can define a function f taking $\mathcal{E} \to \mathcal{O}$ by letting $f(k) = k - 1$, for each k in $\mathcal{E}$. To see that this f is one-to-one, simply note that $k_1 - 1 = k_2 - 1$ implies $k_1 = k_2$. Also, f is clearly onto. Thus, the sets $\mathcal{E}$ and $\mathcal{O}$ have the same cardinality. □

It is not very surprising that the set of even natural numbers and the set of odd natural numbers have the same cardinalities. The following example is a little more unexpected.

Example 10.1.11. The set of even natural numbers has the same cardinality as the set of all natural numbers.

Proof. This is surprising at first because it seems that the set of even numbers should have half as many elements as the set of all natural numbers. However, it is easy to prove that these sets, $\mathcal{E}$ and $\mathbb{N}$, do have the same cardinality. Simply define the function $f : \mathbb{N} \rightarrow \mathcal{E}$ by $f(n) = 2n$. It is easily seen that f is one-to-one: if $f(n_1) = f(n_2)$, then $2n_1 = 2n_2$, so $n_1 = n_2$. The function f is onto since every even number is of the form $2k$, for some natural number k. Therefore, $\mathbb{N}$ and $\mathcal{E}$ have the same cardinality.   □

Thus, in the sense of the definition we are using, the subset $\mathcal{E}$ of $\mathbb{N}$ has the same size as the entire set $\mathbb{N}$. This shows that, with respect to cardinality, it is not necessarily the case that "the whole is greater than any of its parts."

Another example showing that "the whole" can have the same cardinality as "one of its parts" is the following.

Example 10.1.12. The set of natural numbers and the set of nonnegative integers have the same cardinality.

Proof. The set of natural numbers is $\mathbb{N} = \{1, 2, 3, \ldots\}$. Let S denote the set $\{0, 1, 2, 3, \ldots\}$ of nonnegative integers. We want to construct a one-to-one function f taking S onto $\mathbb{N}$. We can simply define f by $f(n) = n + 1$, for each n in S. Clearly f maps S onto $\mathbb{N}$. Also, $f(n_1) = f(n_2)$ implies $n_1 + 1 = n_2 + 1$, which gives $n_1 = n_2$. That is, f does not send two different integers to the same natural number, so f is one-to-one. Therefore, $\mathbb{N}$ and S have the same cardinality.   □

The following notation is useful.

Notation 10.1.13. We use the notation $|S| = |T|$ to mean that S and T have the same cardinality.

Therefore, as shown above, $|\mathcal{O}| = |\mathcal{E}| = |\mathbb{N}|$.

How does the size of the set of all positive rational numbers, which we will denote by $\mathbb{Q}^+$, compare to the size of the set of natural numbers? The subset of $\mathbb{Q}^+$ consisting of those rational numbers with numerator 1 can obviously be paired with $\mathbb{N}$: simply pair $\frac{1}{n}$ with n, for each n in $\mathbb{N}$. But then there are all the rational numbers with numerator 2, and with numerator 3, and so on. It seems that there are many more positive rational numbers than there are natural numbers. However, we now prove that $|\mathbb{N}| = |\mathbb{Q}^+|$.

Theorem 10.1.14. *The set of natural numbers and the set of positive rational numbers have the same cardinality.*

Proof. To prove this theorem, we first describe a way of displaying all the positive rational numbers. We imagine writing all the rational numbers with numerator 1 in one line, and then, underneath that, the rational numbers with numerator 2 in a line, and under that the rational numbers with numerator 3 in a line, and so on. That is, we consider the following array:

$$\frac{1}{1} \quad \frac{1}{2} \quad \frac{1}{3} \quad \frac{1}{4} \quad \frac{1}{5} \quad \frac{1}{6} \quad \frac{1}{7} \quad \cdots$$

$$\frac{2}{1} \quad \frac{2}{2} \quad \frac{2}{3} \quad \frac{2}{4} \quad \frac{2}{5} \quad \frac{2}{6} \quad \frac{2}{7} \quad \cdots$$

$$\frac{3}{1} \quad \frac{3}{2} \quad \frac{3}{3} \quad \frac{3}{4} \quad \frac{3}{5} \quad \frac{3}{6} \quad \frac{3}{7} \quad \cdots$$

$$\frac{4}{1} \quad \frac{4}{2} \quad \frac{4}{3} \quad \frac{4}{4} \quad \frac{4}{5} \quad \frac{4}{6} \quad \frac{4}{7} \quad \cdots$$

$$\vdots \quad \vdots \quad \vdots \quad \vdots \quad \vdots \quad \vdots \quad \vdots$$

Imagining the positive rational numbers arranged as above, we can show that the natural numbers can be paired with them. That is, we will define a one-to-one function f taking $\mathbb{N}$ onto $\mathbb{Q}^+$. As we define the function, you should keep looking back at the array to see the pattern that we are using.

Define $f(1) = \frac{1}{1}$ and $f(2) = \frac{1}{2}$. (We can't continue by $f(3) = \frac{1}{3}$, $f(4) = \frac{1}{4}$, ..., for then f would only map onto those rational numbers with numerator 1.) Define $f(3) = \frac{2}{1}$ and $f(4) = \frac{3}{1}$. We can't just keep going down in our array; we must include the numbers above as well. We need not include $\frac{2}{2}$ however, since $\frac{2}{2} = \frac{1}{1}$, which is already paired with 1. Thus, we let $f(5) = \frac{1}{3}$, $f(6) = \frac{1}{4}$, $f(7) = \frac{2}{3}$, $f(8) = \frac{3}{2}$, $f(9) = \frac{4}{1}$, and $f(10) = \frac{5}{1}$. We need not consider $\frac{4}{2}$, since $\frac{4}{2} = \frac{2}{1}$, and we need not consider $\frac{3}{3} = \frac{1}{1}$ or $\frac{2}{4} = \frac{1}{2}$. Thus, $f(11)$ is defined to be $\frac{1}{5}$ and $f(12) = \frac{1}{6}$. It is apparent that a pairing of the natural numbers and the positive rational numbers is indicated by continuing to label rational numbers with natural numbers in this manner, "zigzagging," you might say, through the above array. Therefore, $|\mathbb{Q}^+| = |\mathbb{N}|$.   □

10.2   Countable Sets and Uncountable Sets

You may be wondering whether or not every infinite set can be paired with the set of natural numbers. If the elements of a set can be paired with the natural numbers, then the elements can be listed in a sequence. For example, if we let s_1 be the element of the set corresponding to the natural number 1, s_2 be the element of the set corresponding to the natural number 2, s_3 to 3, and so on, then the set could be displayed:

$$\{s_1, s_2, s_3, \ldots\}$$

Pairing the elements of a set with the set of natural numbers is, in a sense, "counting the elements of the set."

Definition 10.2.1. A set is *countable* (sometimes called *denumerable*, or *enumerable*) if it is either finite or has the same cardinality as the set of natural numbers. A set is said to be *uncountable* if it is not countable.

One example of an uncountable set is the following.

Theorem 10.2.2. *The set of all real numbers between 0 and 1 is uncountable.*

Proof. We must show that there is no way of pairing the set of natural numbers with the set of real numbers between 0 and 1. Let S denote the set of real numbers between 0 and 1: $S = \{x : 0 \leq x \leq 1\}$. We will show that every pairing of natural numbers with elements of S fails to include some members of S. In other words, we will show that there does not exist any function that maps $\mathbb{N}$ onto S.

Note that the elements of S can be written as infinite decimals; that is, in the form $.c_1c_2c_3\ldots$, where each c_i is a digit between 0 and 9. Some numbers have two different such representations. For example, $.9999\ldots$ is the same number as $1.0000\ldots$, and $.19999\ldots$ is the same number as $.20000\ldots$. For the rest of this proof, let us agree that we choose the representation involving an infinite string of 9's rather than the representation involving an infinite string of 0's for all numbers that have two different representations.

Suppose, then, that f is any function taking $\mathbb{N}$ to S. To show that f cannot be onto, we imagine writing out all the values of f in a list, as follows:

$$f(1) = .a_{11}a_{12}a_{13}a_{14}a_{15}\ldots$$

$$f(2) = .a_{21}a_{22}a_{23}a_{24}a_{25}\ldots$$

$$f(3) = .a_{31}a_{32}a_{33}a_{34}a_{35}\ldots$$

$$f(4) = .a_{41}a_{42}a_{43}a_{44}a_{45}\ldots$$

$$f(5) = .a_{51}a_{52}a_{53}a_{54}a_{55}\ldots$$

$$\vdots$$

We now construct a number in S that is not in the range of the function f. We do that by showing how to choose digits b_j so that the number $x = .b_1b_2b_3b_4\ldots$ is not in the range of f. Begin by choosing $b_1 = 3$ if $a_{11} \neq 3$ and $b_1 = 4$ if $a_{11} = 3$. No matter what digits we choose for the b_j for $j \geq 2$, the number x will be different from $f(1)$ since its first digit is different from the first digit of $f(1)$. Then choose $b_2 = 3$ if $a_{22} \neq 3$ and $b_2 = 4$ if $a_{22} = 3$. This insures that $x \neq f(2)$. We continue in this manner, choosing $b_j = 3$ if $a_{jj} \neq 3$ and $b_j = 4$ if $a_{jj} = 3$, for every natural number j. The number x that is so constructed differs from $f(j)$ in its jth digit. Therefore, $f(j) \neq x$ for all j, so x is not in the range of f. Thus, we have proven that there is no function (one-to-one or otherwise) taking $\mathbb{N}$ onto S, so we conclude that S has cardinality different from that of $\mathbb{N}$.   □

Of course, any given function f in the above proof could be modified so as to produce a function whose range does include the specific number x that we constructed in the course of the proof. For example, given any f, define the function $g : \mathbb{N} \to S$ by defining $g(1) = x$ and $g(n) = f(n - 1)$, for $n \geq 2$. The range of g includes x and also includes the range of f. However, g does not map $\mathbb{N}$ onto S, for the above proof could be used to produce a different x that is not in the range of g.

Definition 10.2.3. For a and b real numbers with $a \leq b$, the *closed interval from a to b* is the set of all real numbers between a and b, including a and b. It is denoted $[a, b]$. That is, $[a, b] = \{x : a \leq x \leq b\}$.

The theorem we have just proven asserts that the closed unit interval, $[0, 1]$, is uncountable. How does the cardinality of other closed intervals compare to that of $[0, 1]$?

Theorem 10.2.4. *If a and b are real numbers and $a < b$, then $[a, b]$ and $[0, 1]$ have the same cardinality.*

Proof. The theorem will be established if we can construct a function $f : [0, 1] \to [a, b]$ that is one-to-one and onto. That is easy to do. Simply define f by $f(x) = a + (b - a)x$. Then $f(0) = a$ and $f(1) = b$. Moreover, the function f increases from a to b as x increases from 0 to 1. If $y \in [a, b]$, let $x = \frac{y-a}{b-a}$. Then $x \in [0, 1]$ and $f(x) = y$. This shows that f is onto. To show that f is one-to-one, assume that $a + (b - a)x_1 = a + (b - a)x_2$. Subtracting a from both sides of this equation and then dividing both sides by $b - a$ yields $x_1 = x_2$. This shows that f is one-to-one. Thus, f is a pairing of the elements of $[0, 1]$ with the elements of $[a, b]$, so $|[0, 1]| = |[a, b]|$. $\qquad\square$

There are other intervals that frequently arise in mathematics.

Definition 10.2.5. If a and b are real numbers and $a < b$, then the *open interval between a and b*, denoted (a, b), is defined by

$$(a, b) = \{x : a < x < b\}$$

The *half-open intervals* are defined by

$$(a, b] = \{x : a < x \leq b\} \text{ and } [a, b) = \{x : a \leq x < b\}$$

How does the size of a half-open interval compare to the size of the corresponding closed interval?

Theorem 10.2.6. *The intervals $[0, 1]$ and $(0, 1]$ have the same cardinality.*

Proof. We want to construct a one-to-one function f taking $[0, 1]$ onto $(0, 1]$. We will define $f(x) = x$ for most x in $[0, 1]$, but we need to make a place for 0 to go to in the half-open interval. For each natural number n, the rational number $\frac{1}{n}$ is in both intervals. Define f on those numbers by $f\left(\frac{1}{n}\right) = \frac{1}{n+1}$, for $n \in \mathbb{N}$. In particular, $f(1) = \frac{1}{2}$. Note that the number 1, which is in $(0, 1]$, is not in the range of f as defined so far. We define $f(0)$ to be 1. We define f on the rest of $[0, 1]$ by $f(x) = x$. That is, $f(x) = x$ for those x other than 0 that are not of the form $\frac{1}{n}$ with n a natural number. It is straightforward to check that we have constructed a one-to-one function mapping $[0, 1]$ onto $(0, 1]$.   □

Suppose that $|\mathcal{S}| = |\mathcal{T}|$ and $|\mathcal{T}| = |\mathcal{U}|$; must $|\mathcal{S}| = |\mathcal{U}|$? If this was not the case, we would be using the "equals" sign in a very peculiar way.

Theorem 10.2.7. *If $|\mathcal{S}| = |\mathcal{T}|$ and $|\mathcal{T}| = |\mathcal{U}|$, then $|\mathcal{S}| = |\mathcal{U}|$.*

Proof. By hypothesis, there exist one-to-one functions f and g mapping $\mathcal{S}$ onto $\mathcal{T}$ and $\mathcal{T}$ onto $\mathcal{U}$, respectively. That is, $f : \mathcal{S} \to \mathcal{T}$ and $g : \mathcal{T} \to \mathcal{U}$. Let $h = g \circ f$ be the composition of g and f. In other words, h is the function defined on $\mathcal{S}$ by $h(s) = g(f(s))$. We must show that h is a one-to-one function taking $\mathcal{S}$ onto $\mathcal{U}$. Let u be any element of $\mathcal{U}$. Since g is onto, there exists a t in $\mathcal{T}$ such that $g(t) = u$. Since f is onto, there is an s in $\mathcal{S}$ such that $f(s) = t$. Then $h(s) = g(f(s)) = g(t) = u$. Thus, h is onto.

To see that h is one-to-one, suppose that $h(s_1) = h(s_2)$; we must show that $s_1 = s_2$. Now $g(f(s_1)) = g(f(s_2))$, so $f(s_1) = f(s_2)$ since g is one-to-one. But f is also one-to-one, and so $s_1 = s_2$. We have shown that h is one-to-one and onto, from which it follows that $|\mathcal{S}| = |\mathcal{U}|$.   □

Theorem 10.2.8. *If a, b, c, and d are real numbers with $a < b$ and $c < d$, then the half-open intervals $(a, b]$ and $(c, d]$ have the same cardinality.*

Proof. The function f defined by $f(x) = a + (b - a)x$ is a one-to-one function mapping $(0, 1]$ onto $(a, b]$, as can be seen by a proof almost exactly the same as that in Theorem 10.2.4. Hence, $\big|(0, 1]\big| = \big|(a, b]\big|$. Similarly the function g defined by $g(x) = c + (d - c)x$ is a one-to-one function mapping $(0, 1]$ onto $(c, d]$, so $\big|(0, 1]\big| = \big|(c, d]\big|$. It follows from Theorem 10.2.7 that $\big|(a, b]\big| = \big|(c, d]\big|$.   □

Are there more positive real numbers than there are numbers in $[0, 1]$? The, perhaps surprising, answer is "no."

Theorem 10.2.9. *The cardinality of the set of nonnegative real numbers is the same as the cardinality of the unit interval $[0, 1]$.*

Proof. We begin by showing that the set $\mathcal{S} = \{x : x \geq 1\}$ has the same cardinality as $(0, 1]$. Note that the function f defined by $f(x) = \frac{1}{x}$ maps $\mathcal{S}$ into $(0, 1]$. For if $x \geq 1$, then $\frac{1}{x} \leq 1$. Also, f maps $\mathcal{S}$ onto $(0, 1]$. For if $y \in (0, 1]$, then $\frac{1}{y} \geq 1$ and $f\left(\frac{1}{y}\right) = y$. To see that f is one-to-one, suppose that $f(x_1) = f(x_2)$. Then $\frac{1}{x_1} = \frac{1}{x_2}$, so $x_1 = x_2$. Hence, f is one-to-one and onto, and it follows that $|\mathcal{S}| = \big|(0, 1]\big|$.

Now let $\mathcal{T} = \{x : x \geq 0\}$. Define the function g by $g(x) = x - 1$. Then g is obviously a one-to-one function mapping $\mathcal{S}$ onto $\mathcal{T}$. Hence $|\mathcal{T}| = |\mathcal{S}|$. Therefore, by Theorem 10.2.7, $|\mathcal{T}| = \big|(0, 1]\big|$. But, by Theorem 10.2.6, $\big|[0, 1]\big| = \big|(0, 1]\big|$. It follows that $|\mathcal{T}| = \big|[0, 1]\big|$.   □

Must the union of two countable sets be countable? A much stronger result is true.

Theorem 10.2.10. *The union of a countable number of countable sets is countable.*

Proof. This can be proven using ideas similar to those used in the proof of the fact that the set of positive rational numbers is countable (see Theorem 10.1.14). Recall that "countable" means either finite or having the same cardinality as $\mathbb{N}$. We will prove this theorem for the cases where all the sets are infinite; you should be able to see how to modify the proof if some or all of the cardinalities are finite.

Suppose, then, that we have a countable collection $\{S_1, S_2, S_3, \ldots\}$ of sets, each of which is itself countably infinite. We can label the elements of S_i so that $S_i = \{a_{i1}, a_{i2}, a_{i3}, \ldots\}$. We can imagine displaying the sets in the following array:

$$S_1 = \{a_{11}, a_{12}, a_{13}, a_{14}, a_{15}, a_{16}, a_{17}, \ldots\}$$

$$S_2 = \{a_{21}, a_{22}, a_{23}, a_{24}, a_{25}, a_{26}, a_{27}, \ldots\}$$

$$S_3 = \{a_{31}, a_{32}, a_{33}, a_{34}, a_{35}, a_{36}, a_{37}, \ldots\}$$

$$S_4 = \{a_{41}, a_{42}, a_{43}, a_{44}, a_{45}, a_{46}, a_{47}, \ldots\}$$

$$\vdots$$

Let S denote the union of the S_i's. To show that S is countable, we show that we can list all of its elements. Proceed as follows. First, list a_{11} and then a_{12}. Then look at a_{21}. It is possible that a_{21} is one of a_{11} or a_{12}, in which case we do not, of course, list it again. If, however, a_{21} is neither a_{11} nor a_{12}, we list it next. Then look at a_{31}; if it is not yet listed, list it next. Then go back up to a_{22}, then a_{13}, and so on. In this way, we "zigzag" through the entire array (as we did in the proof of Theorem 10.1.14) and list all the elements of S. It follows that S is countable. $\quad\square$

10.3 Comparing Cardinalities

When two sets have different cardinalities, the question arises of whether we can say that one set has cardinality that is less than the cardinality of the other set. What should we mean by saying that the cardinality of one set is less than that of another set? It is easiest to begin with a definition of "less than or equal to" for cardinalities.

Definition 10.3.1. If S and T are sets, we say that S *has cardinality less than or equal to the cardinality of* T, and write $|S| \le |T|$, if there is a subset T_0 of T such that $|S| = |T_0|$.

This is equivalent to saying that there is a one-to-one function mapping S into (not necessarily onto) T. For if f is a one-to-one function mapping S onto T_0, we can regard f as a function taking S into T. Conversely, if f is a one-to-one function mapping S into T, and if T_0 is the range of f, then f gives a pairing of S and T_0.

Example 10.3.2. The function $f : \mathbb{N} \to [0, 1]$ defined by $f(n) = \frac{1}{n}$ establishes that $|\mathbb{N}| \leq |[0, 1]|$, since f is one-to-one.

Note that $|S_0| \leq |S|$ whenever S_0 is a subset of S, since the function $f : S_0 \to S$ defined by $f(s) = s$, for each s in S_0, is clearly one-to-one.

We have defined "$\leq$" for cardinalities; how should we define "$<$"? The following definition is very natural.

Definition 10.3.3. We say that S *has cardinality less than that of* T, and write $|S| < |T|$, if $|S| \leq |T|$ and $|S| \neq |T|$.

Example 10.3.4. If $\mathbb{N}$ is the set of natural numbers and $[0, 1]$ is the unit interval, then $|\mathbb{N}| < |[0, 1]|$.

Proof. By Example 10.3.2, $|\mathbb{N}| \leq |[0, 1]|$, and, by Theorem 10.2.2, $|\mathbb{N}| \neq |[0, 1]|$, so the result follows.   □

Thus, in the sense of the definitions we are using, there are more real numbers in the interval $[0, 1]$ than there are natural numbers.

There is a question that immediately arises from the definition of "less than or equal to" for cardinalities: If S and T are sets such that $|S| \leq |T|$ and $|T| \leq |S|$, must $|S| = |T|$? The language suggests that this question should have an affirmative answer, but that language doesn't prove anything. What does this question come down to? We are given the fact that $|S| \leq |T|$. That is equivalent to the existence of a one-to-one function $f : S \to T$. Similarly, $|T| \leq |S|$ gives a one-to-one function $g : T \to S$. To say that $|S| = |T|$ is equivalent to saying that there exists a function $h : S \to T$ that is both one-to-one and onto. The question, therefore, is whether we can show the existence of such a function h from the existence of the functions f and g.

In addition to being important in justifying the above terminology, the following theorem is often very useful in proving that given sets have the same cardinalities.

The Cantor-Bernstein Theorem 10.3.5. *If S and T are sets such that $|S| \leq |T|$ and $|T| \leq |S|$, then $|S| = |T|$.*

Proof. The hypotheses imply that there exist one-to-one functions $f : S \to T$ and $g : T \to S$. We must construct a one-to-one function h that takes S onto T. To do this we will break S up into three subsets and then define h to be the function f on two of those subsets and the function g^{-1} on the third subset.

Consider any element s of S. Such an s may or may not be in the range of g. If it is in the range of g, then there is exactly one element t_0 in T such that $g(t_0) = s$, since g is one-to-one. Call such an element t_0 the "immediate ancestor" of s. Similarly, if t is in T and $f(s_0) = t$ for some s_0 in S, we say that s_0 is the "immediate ancestor" of t. Thus, elements of S have immediate ancestors in T if they are in the range of g, and elements of T have immediate ancestors in S if they are in the range of f. It is possible that some elements do not have any immediate ancestors.

We will say that an immediate ancestor of an immediate ancestor of an element s in S is an "ancestor" of the element s. That is, if s in S has an immediate ancestor t_0

in $\mathcal{T}$ and t_0 has an immediate ancestor s_0 in $\mathcal{S}$, then s_0 is an ancestor of s. Similarly, if t_1 in $\mathcal{T}$ has an immediate ancestor s_1 in $\mathcal{S}$ and s_1 has an immediate ancestor t_2 in $\mathcal{T}$, we say that t_2 is an ancestor of t_1. We continue backwards whenever possible. That is, we start with a given element and then keep on finding immediate ancestors unless and until we reach an element that does not have an immediate ancestor. All the ancestors in such a chain of ancestors are called ancestors of the original element of $\mathcal{S}$ or $\mathcal{T}$.

For each given element that we start with, there are three possibilities. One possibility is that there is no element in the chain of ancestors which itself does not have any ancestor. That is, it could be that we can keep on going back and back and back indefinitely in the ancestry of a given element. Let $\mathcal{S}_\infty$ denote the set of all those elements s in $\mathcal{S}$ for which we can keep on finding ancestors without stopping. Similarly, let $\mathcal{T}_\infty$ denote the set of all t in $\mathcal{T}$ for which we can keep on finding ancestors without stopping. (It might be noted that it is possible that we can keep on finding ancestors indefinitely, but nonetheless there are only a finite number of distinct ancestors. For example, it would be possible that, for some s in $\mathcal{S}$ and t in $\mathcal{T}$, $f(s) = t$ and $g(t) = s$. Then the immediate ancestor of s would be t, the immediate ancestor of t would be s, the immediate ancestor of s would be t, and so on. Thus, there would be no stopping the process of finding ancestors, in spite of the fact that each of s and t has only two distinct ancestors, s and t. In this situation, $s \in \mathcal{S}_\infty$ and $t \in \mathcal{T}_\infty$.)

Those elements of $\mathcal{S}$ and $\mathcal{T}$ that are not in either of $\mathcal{S}_\infty$ or $\mathcal{T}_\infty$ have what might be called "ultimate ancestors." That is, since the chain of ancestors comes to a stop, there is a most distant ancestor. Of course, one possibility is that the element has no ancestors at all, in which case we say that it itself is its ultimate ancestor. The ultimate ancestor of any given element is either in $\mathcal{S}$ or in $\mathcal{T}$. Let $\mathcal{S}_S$ denote the set of all elements of $\mathcal{S}$ whose ultimate ancestor is in $\mathcal{S}$ and let $\mathcal{S}_T$ denote the set of all elements of $\mathcal{S}$ whose ultimate ancestor is in $\mathcal{T}$. Similarly, let $\mathcal{T}_S$ and $\mathcal{T}_T$ denote the sets of elements of $\mathcal{T}$ whose ultimate ancestors are in $\mathcal{S}$ and $\mathcal{T}$, respectively.

Thus, we have divided $\mathcal{S}$ into three subsets: $\mathcal{S}_\infty$, $\mathcal{S}_S$, and $\mathcal{S}_T$. Every element of $\mathcal{S}$ is in exactly one of those subsets. Similarly, every element of $\mathcal{T}$ is in exactly one of the subsets $\mathcal{T}_\infty$, $\mathcal{T}_S$, or $\mathcal{T}_T$. (Of course, some of the subsets may be empty.)

We can now define the function h. For s in $\mathcal{S}$, we define $h(s)$ to be $f(s)$ if s is in either $\mathcal{S}_\infty$ or $\mathcal{S}_S$, and we define $h(s)$ to be $g^{-1}(s)$ if s is in $\mathcal{S}_T$. Note that $g^{-1}(s)$ is defined for all $s \in \mathcal{S}_T$ since all the elements of $\mathcal{S}_T$ have immediate ancestors in $\mathcal{T}$. We will show that h is a one-to-one function taking $\mathcal{S}$ onto $\mathcal{T}$.

Let's first show that h is one-to-one. Suppose that $h(s_1) = h(s_2)$ for s_1 and s_2 in $\mathcal{S}$. We must show that $s_1 = s_2$. If both of s_1 and s_2 are in the union of $\mathcal{S}_\infty$ and $\mathcal{S}_S$, then $h(s_1) = f(s_1)$ and $h(s_2) = f(s_2)$. Therefore, $f(s_1) = f(s_2)$. Since f is one-to-one, it follows that $s_1 = s_2$ in this case. Similarly, if both of s_1 and s_2 are in $\mathcal{S}_T$, then $h(s_1) = g^{-1}(s_1)$ and $h(s_2) = g^{-1}(s_2)$. Therefore, $g^{-1}(s_1) = g^{-1}(s_2)$. Applying g to both sides of this equation gives $s_1 = s_2$ in this case.

One case remains: the case where one of s_1 and s_2 is in the union of $\mathcal{S}_S$ and $\mathcal{S}_\infty$ and the other is in $\mathcal{S}_T$. Suppose that $s_1 \in \mathcal{S}_\infty \cup \mathcal{S}_S$ and $s_2 \in \mathcal{S}_T$. Then $h(s_1) = f(s_1)$ and $h(s_2) = g^{-1}(s_2)$. Therefore, $f(s_1) = g^{-1}(s_2)$. We show that this case cannot arise. If $f(s_1) = g^{-1}(s_2)$, then s_1 is an immediate ancestor of $g^{-1}(s_2)$. Thus, s_1 is an ancestor of s_2. But s_2 is in $\mathcal{S}_T$ so it has an ultimate ancestor in $\mathcal{T}$. Since s_1 is an ancestor of s_2, the ultimate ancestor of s_2 is the ultimate ancestor of s_1. But s_1 being in $\mathcal{S}_\infty \cup \mathcal{S}_S$ implies that s_1 either has no ultimate ancestor or has an ultimate ancestor in $\mathcal{S}$. This is inconsistent with having an ultimate ancestor in $\mathcal{T}$, so this case does not arise.

We have proven that the function h that we constructed is one-to-one. We must now show that h maps $\mathcal{S}$ onto $\mathcal{T}$.

Each t in $\mathcal{T}$ is in one of $\mathcal{T}_S$, $\mathcal{T}_\infty$, or $\mathcal{T}_T$. We must show that, wherever t lies, there is an s in $\mathcal{S}$ such that $h(s) = t$. Suppose first that $t \in \mathcal{T}_S$. Since t has an ultimate ancestor in $\mathcal{S}$, in particular we know that t is in the range of f, so we can consider $f^{-1}(t)$. The ancestors of $f^{-1}(t)$ are also ancestors of t, from which it follows that the ultimate ancestor of $f^{-1}(t)$ is in $\mathcal{S}$. That is, $f^{-1}(t)$ is in $\mathcal{S}_S$. The function h is defined to be f on $\mathcal{S}_S$, so $h(f^{-1}(t)) = f(f^{-1}(t)) = t$. This shows that the range of h contains every element of $\mathcal{T}_S$.

Now consider any t in $\mathcal{T}_\infty$. Such a t has an immediate ancestor in $\mathcal{S}$, $f^{-1}(t)$. Since the ancestors of $f^{-1}(t)$ are also ancestors of t, $f^{-1}(t)$ has no ultimate ancestor. That is, $f^{-1}(t)$ is in $\mathcal{S}_\infty$. The function h was defined to be the function f on $\mathcal{S}_\infty$, so $h(f^{-1}(t)) = f(f^{-1}(t)) = t$. This proves that the range of h contains $\mathcal{T}_\infty$.

All that remains to be shown is that the range of h includes $\mathcal{T}_T$. Suppose, then, that t is in $\mathcal{T}_T$. Let $s = g(t)$. Then t is the immediate ancestor of s. Thus, the ultimate ancestor of t is the ultimate ancestor of s. Since the ultimate ancestor of t is in $\mathcal{T}$, the ultimate ancestor of s is in $\mathcal{T}$. In other words, s is in $\mathcal{S}_T$. On elements of $\mathcal{S}_T$ h is defined to be g^{-1}. Thus, $h(s) = g^{-1}(s)$ and, since $s = g(t)$, $h(s) = g^{-1}(g(t)) = t$. This establishes that the range of h includes $\mathcal{T}_T$.

We have therefore shown that, for every t in $\mathcal{T}$, whatever subset of $\mathcal{T}$ t lies in, there is an s in $\mathcal{S}$ such that $h(s) = t$. This proves that h is onto.

Therefore, h is a one-to-one function mapping $\mathcal{S}$ onto $\mathcal{T}$, and we conclude that $|\mathcal{S}| = |\mathcal{T}|$. $\qquad\qquad\qquad\square$

Corollary 10.3.6. *If $\mathcal{S}$ is a subset of $\mathcal{T}$ and there exists a function $f : \mathcal{T} \to \mathcal{S}$ that is one-to-one, then $\mathcal{S}$ and $\mathcal{T}$ have the same cardinality.*

Proof. Since $\mathcal{S}$ is a subset of $\mathcal{T}$, $|\mathcal{S}| \leq |\mathcal{T}|$. Since there is a one-to-one function mapping $\mathcal{T}$ into $\mathcal{S}$, it follows that $|\mathcal{T}| \leq |\mathcal{S}|$. By the Cantor–Bernstein Theorem (10.3.5), $|\mathcal{S}| = |\mathcal{T}|$. $\qquad\qquad\qquad\square$

The Cantor–Bernstein Theorem can often be used to simplify proofs that given sets have the same cardinalities.

Theorem 10.3.7. *If $a < b$, then $\big|[a,b]\big| = \big|(a,b)\big| = \big|(a,b]\big| = \big|[a,b)\big|.$*

Proof. Clearly, $\big|(a,b)\big| \leq \big|[a,b]\big|$. Note that $\left[a + \frac{b-a}{3}, b - \frac{b-a}{3}\right]$ is contained in (a,b), so $\left|\left[a + \frac{b-a}{3}, b - \frac{b-a}{3}\right]\right| \leq \big|(a,b)\big|$. But by Theorem 10.2.4, $\big|[a,b]\big| = \left|\left[a + \frac{b-a}{3}, b - \frac{b-a}{3}\right]\right|$. Therefore, $\big|[a,b]\big| \leq \big|(a,b)\big|$. So, by the Cantor–Bernstein Theorem (10.3.5), $\big|[a,b]\big| = \big|(a,b)\big|$.

The proofs for the half-open intervals are almost exactly the same as the above proof for the open interval.   □

What is the cardinality of the set of all real numbers?

Theorem 10.3.8. *The cardinality of the set of all real numbers is the same as the cardinality of the unit interval* $[0, 1]$.

Proof. Let $\mathbb{R}$ denote the set of all real numbers. We will "patch together" some of the results that we have already proven to show that $|\mathbb{R}| \leq |[0, 1]|$.

As we have seen, the set of nonnegative real numbers has the same cardinality as $[0, 1]$ (see Theorem 10.2.9). Thus, there exists a one-to-one function f mapping the set of nonnegative real numbers onto $[0, 1]$. The set of negative real numbers obviously has the same cardinality as the set of positive real numbers, as can be seen by using the mapping that takes x to $-x$. The positive real numbers can be mapped in a one-to-one way into $[0, 1]$. Since $|[0, 1]| = |[3, 4]|$ (Theorem 10.2.4), it follows that the positive numbers can be mapped in a one-to-one way into $[3, 4]$. Then, using the equivalence of the positive and negative real numbers, we conclude that there is a function g mapping the negative real numbers into $[3, 4]$. We now define a function h mapping $\mathbb{R}$ into $[0, 1] \cup [3, 4]$ by letting h be f on the nonnegative numbers and g on the negative numbers. Then h is a one-to-one function mapping $\mathbb{R}$ into a subset of $[0, 1] \cup [3, 4]$, which is a subset of $[0, 4]$. It follows that $|\mathbb{R}| \leq |[0, 4]|$. On the other hand, $[0, 4]$ is a subset of $\mathbb{R}$, so $|[0, 4]| \leq |\mathbb{R}|$, and by the Cantor–Bernstein Theorem (10.3.5), $|\mathbb{R}| = |[0, 4]|$. Since $|[0, 4]| = |[0, 1]|$ (Theorem 10.2.4), the theorem follows.   □

There is a theorem that can often be used to give very easy proofs that sets are countable. The next several results form the basis for that theorem.

Theorem 10.3.9. *A subset of a countable set is countable.*

Proof. Let S be a countable set. If S is finite, then the result is clear. If S is infinite, then there exists a one-to-one function, f, mapping the set of natural numbers onto S. Thus, the elements of S can be listed in a sequence, $(f(1), f(2), f(3), f(4), \ldots)$. If S_0 is a subset of S, then the elements of S_0 correspond to some of the elements in the sequence. Therefore, the elements of S_0 can be listed as well, and hence S_0 is either finite or has the same cardinality as $\mathbb{N}$.   □

Corollary 10.3.10. *If S is any set and there exists a one-to-one function mapping S into the set of natural numbers, then S is countable.*

Proof. Let f be a one-to-one function taking S into $\mathbb{N}$. The range of f is some subset T of $\mathbb{N}$. Since f is a one-to-one function taking S onto T, it follows that $|S| = |T|$. By the previous theorem, T is countable, and therefore so is S.   □

Definition 10.3.11. A *finite sequence* of elements of a set S is an ordered collection of elements of S of the form $(s_1, s_2, s_3, \ldots, s_k)$.

For example, one finite sequence of rational numbers is $\left(-\frac{1}{2}, -7, \frac{22}{7}, 0\right)$.

Theorem 10.3.12. *The set of all finite sequences of natural numbers is countable.*

Proof. Let $\mathbb{N}$ denote the set of natural numbers and let S denote the set of all finite sequences of natural numbers. By the above corollary (10.3.10), it suffices to show that there is a one-to-one function g mapping S into $\mathbb{N}$. Here is a description of one such function. We define the value of g at each given finite sequence of natural numbers to be the number whose digits are 1's and 0's, determined as follows: begin with the number of 1's equal to the first number in the given finite sequence, follow that by a 0, then follow that by the number of 1's equal to the second number in the sequence, then another 0, then the number of 1's corresponding to the third number in the sequence, then a 0, and so on, ending with the number of 1's corresponding to the last number in the sequence. For example,

$$g\big((2,3,7)\big) = 11011101111111$$

and

$$g\big((5,1)\big) = 1111101$$

The function g is one-to-one since you can recover the unique sequence corresponding to any number in the range of g by using the definition of g. For example, the number 1111010111111011111111 corresponds to the sequence $(4,1,6,8)$. Since g is one-to-one and maps S into $\mathbb{N}$, S is countable.   □

Corollary 10.3.13. *If $\mathcal{L}$ is any countable set, then the set of all finite sequences of elements of $\mathcal{L}$ is countable.*

Proof. This follows easily from the above theorem. By hypothesis, there exists a one-to-one function f mapping $\mathcal{L}$ into $\mathbb{N}$. Then, a one-to-one function F mapping sequences of elements of $\mathcal{L}$ into sequences of elements of $\mathbb{N}$ can be obtained by defining

$$F(a_1, a_2, a_3, \ldots, a_k) = \big(f(a_1), f(a_2), f(a_3), \ldots, f(a_k)\big)$$

Thus, the previous theorem implies the corollary.   □

The following definition will be useful.

Definition 10.3.14. Let S and $\mathcal{T}$ be any sets. We will say that $\mathcal{T}$ *can be labeled* by the set S if there is a way of assigning a finite sequence of elements of S to each element of $\mathcal{T}$ so that each finite sequence corresponds to at most one element of $\mathcal{T}$.

Example 10.3.15. The set $\mathbb{Q}$ of rational numbers can be labeled by the set

$$\mathcal{L} = \{0, 1, 2, 3, 4, 5, 6, 7, 8, 9, -, /\}$$

To label a given rational number, use the representation of that rational number in lowest terms and simply write it in the usual way using symbols from the set $\mathcal{L}$.

The following theorem is useful in many situations. It is a slight variant of the "Typewriter Principle" that was developed by Bjorn Poonen.

The Enumeration Principle 10.3.16. *Every set that can be labeled by a countable set is countable.*

Proof. Let $\mathcal{T}$ be a set that is labeled by a countable set $\mathcal{L}$. The fact that no two elements of $\mathcal{T}$ have the same label implies that there is a one-to-one function f mapping $\mathcal{T}$ into the set of labels. Thus, there is a one-to-one function mapping $\mathcal{T}$ into the set of finite sequences of elements of $\mathcal{L}$, which is a countable set by Corollary 10.3.13. It follows from Corollary 10.3.10 that $\mathcal{T}$ is countable. □

Any set that is proven to be countable by the Enumeration Principle could, of course, also be proven to be countable without using this principle. However, the Enumeration Principle often leads to very simple proofs.

Theorem 10.3.17. *The set of all rational numbers is countable.*

Proof. As indicated above (Example 10.3.15), the set of rational numbers can be labeled by the set $\mathcal{L} = \{0, 1, 2, 3, 4, 5, 6, 7, 8, 9, -, /\}$. The Enumeration Principle (10.3.16) gives the result. □

Corollary 10.3.18. *The set of integers is countable.*

Proof. A subset of a countable set is countable (Theorem 10.3.9), so this follows from the previous theorem (10.3.17). □

You may have heard the assertion that π is a *transcendental number*; what does that mean?

Definition 10.3.19. The real number x_0 is said to be *algebraic* if it is the root of a polynomial with integer coefficients. The real number x_0 is said to be *transcendental* if there is no polynomial with integer coefficients that has x_0 as a root.

For example, the number $-\frac{3}{4}$ is algebraic, since it is a root of the polynomial $4x + 3$. More generally, every rational number $\frac{m}{n}$ is algebraic since it is a root of the polynomial $nx - m$. There are also many irrational numbers that are algebraic, such as $\sqrt{2}$ which is a root of the polynomial $x^2 - 2$ and $(\frac{3}{4})^{\frac{1}{5}}$ which is a root of the polynomial $4x^5 - 3$.

It is not so easy to prove the existence of transcendental numbers. It is true that π is transcendental, but it is very difficult to prove that fact. It is somewhat easier, but still quite difficult, to prove that e, the base for natural logarithms, is transcendental. It is a very surprising and beautiful fact that it is much easier to prove that most real numbers are transcendental than it is to prove that any specific real number is transcendental. This is a corollary of the following.

Theorem 10.3.20. *The set of algebraic numbers is countable.*

Proof. We show that the set of algebraic numbers can be labeled by the set of integers and a comma; the Enumeration Principle (10.3.16) then gives the result.

Let x_0 be an algebraic number. We can label x_0 as follows. Specify the degree, n, of a polynomial of least degree with integer coefficients that has x_0 as a root. Then put a comma. If a polynomial of degree n has x_0 as a root, divide the polynomial by the coefficient of x^n to get a polynomial with leading coefficient 1 that has x_0 as a root. Choose one such polynomial. (In fact, there is only one such polynomial; see Problem 35 at the end of this chapter.) Then specify the coefficients of that polynomial in the order corresponding to descending powers of x, each separated by commas. Then put the integer 1, 2, or 3, and so on, to indicate that the algebraic number is the smallest, next to smallest, third from smallest, and so on, of the roots of that polynomial. In this manner, we label every algebraic number by a finite sequence of integers. Since the set of integers is countable (Corollary 10.3.18), it follows from the Enumeration Principle that the set of algebraic numbers is countable.   □

The above easily establishes the existence of transcendental numbers.

Corollary 10.3.21. *There exist transcendental numbers.*

Proof. Since the set of algebraic numbers is countable (Theorem 10.3.20) and the set of all real numbers is uncountable (Theorem 10.3.8 and Theorem 10.2.2), there are some real numbers that are not algebraic and are therefore transcendental.   □

The cardinality of a finite set consisting of n elements is said to be n. We now introduce some standard notation for the sizes of some of the most common infinite sets.

Definition 10.3.22. We say that the set S has *cardinality* $\aleph_0$ (which we read "aleph nought") if the cardinality of S is the same as that of the natural numbers, in which case we write $|S| = \aleph_0$.

For example, $|\mathbb{Q}| = \aleph_0$.

There is also a standard notation for the cardinality of the set of real numbers.

Definition 10.3.23. We say that the set S has *cardinality* c if the cardinality of S is the same as the cardinality of the set of real numbers; c is sometimes said to be the *cardinality of the continuum*.

For example, $|[3, 9]| = c$.

Note that $\aleph_0 < c$, in the sense that every set with cardinality $\aleph_0$ has cardinality less than every set with cardinality c.

It is important to note that $\aleph_0$ is the smallest infinite cardinality in the following sense.

Theorem 10.3.24. *If S is an infinite set, then $\aleph_0 \leq |S|$.*

Proof. To establish this, we must show that S has a subset S_0 of cardinality $\aleph_0$. We proceed as follows. Since S is infinite, it surely contains some element, say s_1. Similarly, $S \setminus \{s_1\}$ (i.e., the set obtained from S by removing s_1) contains some element, say s_2. Similarly, $S \setminus \{s_1, s_2\}$ contains some element s_3. Proceeding in this manner creates an infinite sequence $(s_1, s_2, s_3, \dots)$ of elements of S. Let $S_0 =$

$\{s_1, s_2, s_3, \ldots\}$. Then clearly $|S_0| = |\mathbb{N}| = \aleph_0$. Since S_0 is a subset of S, it follows that $\aleph_0 \leq |S|$.   □

Thus, $\aleph_0$ is the smallest infinite cardinality. Is there a largest cardinality?

Definition 10.3.25. If S is any set, then the set of all subsets of S is called the *power set of S* and is denoted $\mathcal{P}(S)$.

The terminology "power set of S" comes from the following theorem. (This was stated as Problem 5 in Chapter 2.)

Theorem 10.3.26. *If S is a finite set with n elements, then the cardinality of $\mathcal{P}(S)$ is 2^n.*

Proof. First note that this is true for $n = 0$. For the only set with 0 elements is $\emptyset$, the empty set. The empty set has one subset, namely itself. Since $2^0 = 1$, the theorem holds for $n = 0$.

We proceed by mathematical induction. Suppose that every set with k elements has 2^k subsets and let S be a set with $k + 1$ elements. Suppose that s_0 is any element of S and let S_0 be the subset $S \setminus \{s_0\}$ of S obtained by removing s_0. Then S_0 has k elements and, by the inductive hypothesis, $|\mathcal{P}(S_0)| = 2^k$. Suppose that T is any subset of S_0. Then T is also a subset of S. The set $T \cup \{s_0\}$ is a different subset of S. Thus, for each subset T of S_0, there are two subsets of S, T and $T \cup \{s_0\}$. It follows that there are twice as many subsets of S as there are of S_0. That is,

$$|\mathcal{P}(S)| = 2 \cdot |\mathcal{P}(S_0)| = 2 \cdot 2^k = 2^{k+1}$$

The theorem follows by mathematical induction.   □

What is the relationship between $|S|$ and $|\mathcal{P}(S)|$ when S is an infinite set?

Theorem 10.3.27. *For every set S, $|S| < |\mathcal{P}(S)|$.*

Proof. It is easy to see that $|S| \leq |\mathcal{P}(S)|$. Among the subsets of S are the "singleton sets;" i.e., sets of the form $\{s\}$, for each $s \in S$. The collection $\mathcal{P}_0$ of all singleton subsets of S is a subset of $\mathcal{P}(S)$. A one-to-one function f mapping S into $\mathcal{P}(S)$ can be defined by $f(s) = \{s\}$, for all s in S. Thus, $|S| = |\mathcal{P}_0|$, so $|S| \leq |\mathcal{P}(S)|$.

To show that $|S| < |\mathcal{P}(S)|$, we must show that there is no one-to-one function f taking S onto $\mathcal{P}(S)$. The proof will use a "diagonal argument" similar to the proof we gave that $[0, 1]$ is uncountable (Theorem 10.2.2).

Suppose, then, that f is any function taking S into $\mathcal{P}(S)$. We will show that f cannot be onto; that is, that there is an element of $\mathcal{P}(S)$ (i.e., a subset of S) that is not in the range of f.

For each $s \in S$, $f(s)$ is a subset of S. Define the subset S_0 of S by

$$S_0 = \{s \in S : s \notin f(s)\}$$

That is, the subset S_0 of S is defined to consist of all of those elements s of S that are not in the subset of S that f assigns to s.

The set S_0 is an element of $\mathcal{P}(S)$. We will show that it is not in the range of f. To prove this by contradiction, suppose that there was some $s_0 \in S$ such that $f(s_0) = S_0$. We show that this is impossible by asking the question: is s_0 in S_0? We will see that this question does not have an answer. Suppose that $s_0 \notin S_0$. The definition of S_0 is that it contains those elements of S that are not in the subsets they are sent to by f. Thus, if s_0 is not in $f(s_0)$, s_0 is in S_0. In other words, $s_0 \notin S_0$ implies $s_0 \in S_0$, which is a contradiction.

On the other hand, if s_0 is in S_0, then the definition of S_0 implies that s_0 is not in $f(s_0)$, but since $f(s_0) = S_0$, $s_0 \notin S_0$. Thus, $s_0 \in S_0$ implies $s_0 \notin S_0$, which is also a contradiction. But if there was an s_0 satisfying $f(s_0) = S_0$, then s_0 would either be in S_0 or not be in S_0. Therefore, there is no s_0 satisfying $f(s_0) = S_0$, and the theorem is proven.   □

One of the consequences of the theorem we have just established is that there is no largest cardinal number. For if S is any set, there is a set whose cardinality is bigger than that of S, namely $\mathcal{P}(S)$.

In particular, for the set of real numbers $\mathbb{R}$, the cardinality of $\mathcal{P}(\mathbb{R})$, the set of all sets of real numbers, is greater than c. Because of the analogy to the case of finite sets, it is standard to write $|\mathcal{P}(\mathbb{R})| = 2^c$.

Similarly, $2^{\aleph_0}$ denotes the cardinality of $\mathcal{P}(\mathbb{N})$. By the above, $\aleph_0 < 2^{\aleph_0}$. Also, as we have seen, $\aleph_0 < c$. What is the relationship between $2^{\aleph_0}$ and c?

Theorem 10.3.28. *The cardinality of the set of all sets of natural numbers is the same as the cardinality of the set of real numbers. That is,* $|\mathcal{P}(\mathbb{N})| = c$, *or* $2^{\aleph_0} = c$.

Proof. Since $\big| [0, 1] \big| = |\mathbb{R}|$ (Theorem 10.3.8), it suffices to prove that $\big| [0, 1] \big| = |\mathcal{P}(\mathbb{N})|$. It will be convenient to introduce another set. Let S denote the set of all infinite sequences of 0's and 1's (typical elements of S are $\{1, 0, 1, 0, 1, 0, \ldots\}$, $\{1, 1, 0, 1, 1, 1, \ldots\}$, and so on). We begin by showing that $|S| = |\mathcal{P}(\mathbb{N})|$. For this, we define a function f taking S into $\mathcal{P}(\mathbb{N})$ by

$$f\big(\{a_1, a_2, a_3, \ldots\}\big) = \{i : a_i = 1\}$$

That is, f takes a sequence of 0's and 1's to the set of those natural numbers consisting of the places where the sequence has 1's. It is clear that f is one-to-one, for two different sequences would have at least one place where one has a 0 and the other has a 1, and the number corresponding to that place would be in the subset corresponding to the second sequence, but not the first. The function f is also onto, for if $\mathcal{T}$ is any subset of $\mathbb{N}$, define a sequence $\{a_i\}$ by letting $a_i = 1$ if i is in $\mathcal{T}$ and $a_i = 0$ if i is not in $\mathcal{T}$. Then $f\big(\{a_i\}\big) = \mathcal{T}$. Thus, $|S| = |\mathcal{P}(\mathbb{N})|$ and the theorem will be proven if we establish that $|S| = \big| [0, 1] \big|$.

We define the function g mapping S into $[0, 1]$ by

$$g\big(\{a_1, a_2, a_3, \ldots\}\big) = .a_1 a_2 a_3 \ldots$$

Since the a_i's are 0's and 1's, the range of g is contained in $[0, 1]$. Since two different sequences of 0's and 1's are sent by g to two different numbers, g is one-to-one. Thus, $|S| \leq |[0, 1]|$.

For the reverse inequality, we must produce a one-to-one function h that takes $[0, 1]$ into S. To do that, we represent the elements of $[0, 1]$ as "binary decimals." That is, every element of $[0, 1]$ can be written as an infinite sum:

$$\frac{a_1}{2} + \frac{a_2}{2^2} + \frac{a_3}{2^3} + \frac{a_4}{2^4} + \cdots$$

where each a_i is 0 or 1. (Of course, as with ordinary decimal representation, some elements of $[0, 1]$ have more than one such representation. For example, $\frac{1}{2} + \frac{0}{2^2} + \frac{0}{2^3} + \frac{0}{2^4} \cdots$ represents the same number as $\frac{0}{2} + \frac{1}{2^2} + \frac{1}{2^3} + \frac{1}{2^4} \cdots$. In such ambiguous cases, choose either representation.)

After representing the elements of $[0, 1]$ as binary decimals as above, define the mapping h taking $[0, 1]$ into S by

$$h\left(\frac{a_1}{2} + \frac{a_2}{2^2} + \frac{a_3}{2^3} + \frac{a_4}{2^4} + \cdots\right) = \{a_1, a_2, a_3, a_4, \ldots\}$$

The function h is a one-to-one mapping of $[0, 1]$ into S, so $|[0, 1]| \leq |S|$. By the Cantor–Bernstein Theorem (10.3.5), $|[0, 1]| = |S|$, proving the theorem. ☐

Definition 10.3.29. The *unit square in the plane* is the subset of the plane consisting of all points whose x and y coordinates are both between 0 and 1. That is, the unit square is the set S defined by

$$S = \{(x, y) : 0 \leq x \leq 1, 0 \leq y \leq 1\}$$

Theorem 10.3.30. *The cardinality of the unit square in the plane is c.*

Proof. Let S denote the unit square. It is clear that $|S| \geq c$, since S contains the subset

$$S_0 = \{(x, 0) : 0 \leq x \leq 1\}$$

and there is an obvious pairing of S_0 with $[0, 1]$.

To establish the reverse inequality, we will construct a one-to-one function f mapping S into $[0, 1]$. We represent the coordinates of points in the unit square as infinite decimals. In ambiguous cases (i.e., where a representation of a number could end in either a string of 0's or a string of 9's), we choose the representation ending in a string of 9's. We then define the function f by

$$f\left((.a_1 a_2 a_3 \ldots, .b_1 b_2 b_3 \ldots)\right) = .a_1 b_1 a_2 b_2 a_3 b_3 \ldots$$

We claim that f is one-to-one. This follows since $f\big((x, y)\big) = .c_1c_2c_3\ldots$ implies that $x = .c_1c_3c_5\ldots$ and $y = .c_2c_4c_6\ldots$. Thus, $|\mathcal{S}| \leq \big|[0, 1]\big|$, so the Cantor–Bernstein Theorem (10.3.5) gives $|\mathcal{S}| = \big|[0, 1]\big|$.   □

It can be interesting to determine the cardinality of various sets of functions. We present one example now; many other examples are given in the problems. The following definition will be useful.

Definition 10.3.31. If $\mathcal{S}$ is a set and $\mathcal{S}_0$ is a subset of $\mathcal{S}$, then the *characteristic function* of $\mathcal{S}_0$ as a subset of $\mathcal{S}$ is the function f, with domain $\mathcal{S}$, defined by $f(s) = 1$ if $s \in \mathcal{S}_0$ and $f(s) = 0$ if $s \notin \mathcal{S}_0$.

The following is a very easy, but very useful, fact.

Theorem 10.3.32. *For any set $\mathcal{S}$, the set of all characteristic functions with domain $\mathcal{S}$ has the same cardinality as $\mathcal{P}(\mathcal{S})$.*

Proof. As indicated in the definition above of characteristic function, each subset does have a characteristic function. On the other hand, if two characteristic functions are equal as functions, they must be characteristic functions of the same subset (the subset consisting of all elements of the set on which the functions have value 1). Thus, the correspondence between the set of subsets of $\mathcal{S}$ and characteristic functions with domain $\mathcal{S}$ is one-to-one and onto.   □

Theorem 10.3.33. *The cardinality of the set of all functions mapping $[0, 1]$ into $[0, 1]$ is 2^c.*

Proof. Among the functions are those that take on values contained in $\{0, 1\}$; i.e., the characteristic functions with domain $[0, 1]$. By the previous theorem (10.3.32), this set of characteristic functions has cardinality 2^c. Thus, the set of all functions mapping $[0, 1]$ into $[0, 1]$ has cardinality at least 2^c.

To prove the reverse inequality, recall that every function is determined by its graph. The graph of a function f from $[0, 1]$ to $[0, 1]$ is $\{(x, f(x)) : x \in [0, 1]\}$, which is a subset of the unit square. Thus, the set of functions we are considering corresponds to a collection of some of the subsets of the unit square and hence has cardinality at most equal to that of the set of all subsets of the unit square. We have seen (Theorem 10.3.30) that the cardinality of the unit square is c. It follows that the cardinality of the set of *all* subsets of the unit square is 2^c. Therefore, the cardinality of the set of graphs of functions is at most 2^c. By the Cantor–Bernstein Theorem (10.3.5), the cardinality of the set of functions is 2^c.   □

There are some serious deficiencies in the general approach to set theory that we have been describing. The following illustrates some of the problems.

Cantor's Paradox 10.3.34. Let $\mathcal{S}$ denote the set of all sets. Then every subset of $\mathcal{S}$ is an element of $\mathcal{S}$, since each subset is a set. That is, $\mathcal{P}(\mathcal{S})$ is a subset of $\mathcal{S}$. Hence, $|\mathcal{P}(\mathcal{S})| \leq |\mathcal{S}|$. On the other hand, $|\mathcal{S}| < |\mathcal{P}(\mathcal{S})|$ (by Theorem 10.3.27). The Cantor–Bernstein Theorem (10.3.5) proves that this is a contradiction.

What does this contradiction mean? If there is a contradiction, then something is false; but what? The only assumption that we have made is that there *is* a set consisting of the set of all sets. This contradiction shows that there cannot be such a set. To avoid Cantor's Paradox, the definition of set has to be more restrictive.

There is another paradox similar to Cantor's.

Russell's Paradox 10.3.35. Define a set to be *ordinary* if it is not an element of itself. (That is, $S \notin S$.) All of the sets that we have discussed so far except for the set of all sets are ordinary sets. Each set is, of course, a *subset* of itself, but that is very different from being a member of itself. (For example, the set of natural numbers is not a natural number.)

Let $\mathcal{T}$ denote the set of all ordinary sets. We now ask the question: is $\mathcal{T}$ an ordinary set? If $\mathcal{T}$ was an ordinary set, then, since $\mathcal{T}$ is the set of all ordinary sets, $\mathcal{T} \in \mathcal{T}$. But then $\mathcal{T}$ would not be an ordinary set, since it would be an element of itself. On the other hand, if $\mathcal{T}$ is not an ordinary set, then $\mathcal{T} \in \mathcal{T}$. But every element of $\mathcal{T}$ is an ordinary set, so it would follow that $\mathcal{T}$ is ordinary. That is, if $\mathcal{T}$ is ordinary, it is not ordinary; if $\mathcal{T}$ is not ordinary, it is ordinary. There cannot be such a set.

Note that the Cantor and Russell paradoxes are related in the following sense: the set of all sets, if it existed, would be a set that is not an ordinary set.

When mathematicians first encountered the Cantor and Russell paradoxes, over a hundred years ago, they were very concerned. Why aren't "the set of all sets" and "the set of all ordinary sets" themselves sets? What other "sets" are not really sets?

The above and related paradoxes do not arise when considering the sets that naturally arise in doing mathematics. Mathematicians have developed several different "axiomatic set theories" in which the concept of "set" is restricted so that the Cantor and Russell paradoxes do not arise. In these set theories, there are no sets that are elements of themselves. The most popular of the axiomatic set theories is called *Zermelo–Fraenkel Set Theory*. The development of axiomatic set theories is fairly complicated and we will not discuss it here. However, the theorems that we presented in this chapter are also theorems in Zermelo–Fraenkel Set Theory although the formal proofs are slightly different.

The following is a very natural question: is there any set S whose cardinality is greater than $\aleph_0$ and less than c? If there is such a set, it could be mapped into $\mathbb{R}$. That is, if there is any such set, there is a subset of $\mathbb{R}$ with that property. The question can therefore be reformulated: if S is an uncountable subset of $\mathbb{R}$, must the cardinality of S be c? This appears to be a very concrete question. It can be made even more concrete, as follows: if S is a subset of $\mathbb{R}$ and there is no one-to-one function taking S into $\mathbb{N}$, must there exist a one-to-one function taking S onto $\mathbb{R}$?

The Continuum Hypothesis 10.3.36. *There is no set with cardinality strictly between $\aleph_0$ and c.*

It is very surprising that it is not known whether the Continuum Hypothesis is true or false. It is even more surprising that it has been proven that the Continuum Hypothesis is an undecidable proposition in the following sense: it has been established that the Continuum Hypothesis can neither be proven nor disproven within standard set theories, such as Zermelo–Fraenkel Set Theory. Mathematicians disagree about the full implications of this. It is our view that it is possible that someone will prove the Continuum Hypothesis in a way that would convince all mathematicians, in spite of its being undecidable within Zermelo–Fraenkel Set Theory. That is, someone might begin a proof as follows: "Let S be an uncountable subset of $\mathbb{R}$. We construct a one-to-one function f mapping S onto $\mathbb{R}$ by first ...." Any such proof would have to use something that was not part of Zermelo–Fraenkel Set Theory, since it has been proven that the Continuum Hypothesis cannot be decided within Zermelo–Fraenkel Set Theory. On the other hand, it is our opinion that it is possible that a proof could be found that would be based on properties of the set of real numbers that virtually every mathematician would agree are true, in spite of the fact that at least one of them would not be part of Zermelo–Fraenkel Set Theory. However, many mathematicians believe that Zermelo–Fraenkel Set Theory captures all the reasonable properties of the real numbers and therefore conclude that no such proof is possible. We invite you to try to prove that those mathematicians are wrong by proving (or disproving) the Continuum Hypothesis.

10.4   Problems

Basic Exercises

1. Show that the set of all polynomials with rational coefficients is countable.
2. Suppose that the sets S, T, and U satisfy $S \subset T \subset U$ and that $|S| = |U|$. Show that T has the same cardinality as S.
3. Let A and B be countable sets. Prove that the *Cartesian product* of A and B, $A \times B = \{(a, b) : a \in A, b \in B\}$, is countable.
4. Assume that $|A_1| = |B_1|$ and $|A_2| = |B_2|$. Prove:

   (a) $|A_1 \times A_2| = |B_1 \times B_2|$.
   (b) If A_1 is disjoint from A_2 and B_1 is disjoint from B_2, then $|A_1 \cup A_2| = |B_1 \cup B_2|$.

5. Prove that the half-open intervals $[0, 1)$ and $(0, 1]$ have the same cardinality. (This was stated but not proven in Theorem 10.3.7.)
6. What is the cardinality of the set of all functions from $\mathbb{N}$ to $\{1, 2\}$?
7. What is the cardinality of the set of all numbers in the interval $[0, 1]$ which have decimal expansions with a finite number of nonzero digits?
8. Let $\mathbb{Q}(\sqrt{2})$ be the set of real numbers of the form $a + b\sqrt{2}$, where a and b are rational numbers. Find the cardinality of $\mathbb{Q}(\sqrt{2})$.

Interesting Problems

9. Suppose that S and T each have cardinality c. Show that $S \cup T$ also has cardinality c.

10. What is the cardinality of $\mathbb{R}^2 = \{(x, y) : x, y \in \mathbb{R}\}$ (the *Euclidean plane*)?

11. What is the cardinality of the set of all complex numbers?

12. Prove that the set of all finite subsets of $\mathbb{Q}$ is countable.

13. Let S and T be finite sets and let $C = \{f : S \to T\}$ be the set of all functions from S to T. Show that if $|T| > 1$, then $|C| \geq 2^{|S|}$.

14. What is the cardinality of the unit cube, where the unit cube is $\{(x, y, z) : x, y, z \in [0, 1]\}$?

15. What is the cardinality of $\mathbb{R}^3 = \{(x, y, z) : x, y, z \in \mathbb{R}\}$?

16. What is the cardinality of the set of all functions from $\{1, 2\}$ to $\mathbb{N}$?

17. Find the cardinality of the set of all points in $\mathbb{R}^3$ all of whose coordinates are rational.

18. Let S be the set of all functions mapping the set $\{\sqrt{2}, \sqrt{3}, \sqrt{5}, \sqrt{7}\}$ into $\mathbb{Q}$. What is the cardinality of S?

19. Find the cardinality of the set $\{(x, y) : x \in \mathbb{R}, y \in \mathbb{Q}\}$.

20. What is the cardinality of the set of all numbers in the interval $[0, 1]$ that have decimal expansions that end with an infinite sequence of 7's?

21. Let t be a transcendental number. Prove that $t^4 + 7t + 2$ is also transcendental.

22. Suppose that T is an infinite set and S is a countable set. Show that $S \cup T$ has the same cardinality as T.

23. Let S be the set of real numbers t such that $\cos t$ is algebraic. Prove that S is countably infinite.

24. Let a, b, and c be distinct real numbers. Find the cardinality of the set of all functions mapping $\{a, b, c\}$ into the set of real numbers.

25. What is the cardinality of

$$\left\{ n^{\frac{1}{k}} : n, k \in \mathbb{N} \right\}$$

i.e., the set of all roots of all natural numbers?

26. Prove that there does not exist a set with a countably infinite power set.

27. (This problem requires some basic facts about trigonometric functions.) Find a one-to-one function mapping the interval $(-\frac{\pi}{2}, \frac{\pi}{2})$ onto $\mathbb{R}$.

Challenging Problems

28.(a) Prove directly that the cardinality of the closed interval $[0, 1]$ is equal to the cardinality of the open interval $(0, 1)$ by constructing a function $f : [0, 1] \to (0, 1)$ that is one-to-one and onto.

(b) More generally, show that if S is an infinite set and $\{a,b\} \subset S$, then $|S| = |S \setminus \{a,b\}|$. (The notation $S \setminus \{a,b\}$ is used to denote the set of all s in S such that s is not in $\{a,b\}$.)

[Hint: Use the fact that S has a countably infinite subset containing a and b.]

29. Prove that a set is infinite if and only if it has the same cardinality as a proper subset (i.e. a subset other than the set itself) of itself.

30. Call a complex number *complex-algebraic* if it is a root of a polynomial with integer coefficients. Prove that the set of all complex–algebraic numbers is countable.

31. What is the cardinality of the set of all finite subsets of $\mathbb{R}$?

32. What is the cardinality of the set of all countable sets of real numbers?

33. Find the cardinality of the set of all lines in the plane.

34. Show that the set of all functions mapping $\mathbb{R} \times \mathbb{R}$ into $\mathbb{Q}$ has cardinality 2^c.

35. Prove the following: If n is the smallest natural number such that a polynomial of degree n with integer coefficients has x_0 as a root, and if p and q are polynomials of degree n with integer coefficients that have the same leading coefficients (i.e., coefficients of x^n) and each have x_0 as a root, then $p = q$.

36. Let S be the set of all real numbers that have a decimal representation using only the digits 2 and 6. Show that the cardinality of S is c.

37. Let S denote the collection of all circles in the plane. Is the cardinality of S equal to c or 2^c?

38. Prove that if S is uncountable and T is a countable subset of S, then the cardinality of $S \setminus T$ (where $S \setminus T$ denotes the set of all elements of S that are not in T) is the same as the cardinality of S.

39. Find the cardinality of the set of all polynomials with real coefficients; that is, of the set of all expressions of the form

$$a_n x^n + a_{n-1} x^{n-1} + \cdots + a_1 x + a_0$$

where n is a nonnegative integer (that depends on the expression) and $a_0, a_1, \ldots, a_n$ are real numbers.

40. Prove that the union of c sets that each have cardinality c has cardinality c.

41. Prove that the set of all sequences of real numbers has cardinality c.

Chapter 11
Fundamentals of Euclidean Plane Geometry

In this chapter we describe the fundamentals of Euclidean geometry of the plane in a way that relies on some intuitively apparent properties of geometric figures. In particular, some of our proofs are based on what is apparent from looking at diagrams. Of course, we also assume various axioms, such as between any two points there exists a unique line that extends infinitely in two directions. Later on in the chapter we will need to make an additional, very important, assumption known as the Parallel Postulate. More rigorous axiomatic approaches to Euclidean geometry are possible.

11.1 Triangles

One basic concept is that of a *triangle*, by which we mean a geometric figure consisting of three points (called its *vertices*) that do not all lie on one line and the line segments joining those vertices (which are called the *sides* of the triangle). Thus, a typical triangle is pictured in Figure 11.1, where its vertices are labeled with capital letters. We often refer to the sides of the triangle as AB (or BA), BC, and AC.

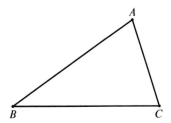

Fig. 11.1 A typical triangle

D. Rosenthal et al., *A Readable Introduction to Real Mathematics*,
Undergraduate Texts in Mathematics, DOI 10.1007/978-3-319-05654-8_11,
© Springer International Publishing Switzerland 2014

The triangle in Figure 11.1 might be denoted $\triangle ABC$. One of the angles might be denoted either $\angle A$ or $\angle BAC$. When we say that two line segments are equal we mean that they have the same length. When we say that two angles are equal we mean that they have the same measure (i.e., one could be placed on top of the other so that they coincide).

Definition 11.1.1. Two triangles are *congruent*, denoted $\cong$, if their vertices can be paired so that the corresponding angles and sides are equal to each other. That is, $\triangle ABC \cong \triangle DEF$ if $\angle A = \angle D$, $\angle B = \angle E$, and $\angle C = \angle F$ and $AB = DE$, $BC = EF$, and $AC = DF$.

If two triangles are congruent, then one can be placed on top of the other so that they completely coincide. More generally, two geometric figures are said to be *congruent* to each other if they can be so placed. It is important to note that congruence of triangles can be established without verifying that all of the pairs of corresponding angles and all the pairs of corresponding sides are equal to each other; as we shall see, equality of certain collections of those pairs implies equality of all of them.

For example, suppose that we fix a side of a triangle and an angle with vertex one of the endpoints of that side and the length of the next side. That is, for example, suppose in Figure 11.2, we fix the angle B and lengths AB and BC. It seems intuitively clear that any two triangles with the specified sides AB and BC and the angle B between them are congruent to each other; the only way to complete the given data to form a triangle is by joining A to C by a line segment. Thus, it appears that any two triangles that have two pairs of equal sides and have equal angles formed by those sides are congruent to each other. We state this as a fundamental axiom.

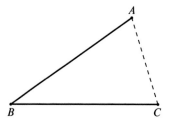

Fig. 11.2 Illustrating side-angle-side

The Congruence Axiom 11.1.2 (Side-Angle-Side). If two triangles have two pairs of corresponding sides equal and also have equal angles between those two sides, then the triangles are congruent to each other.

We speak of this axiom as stating that triangles are congruent if they have "side-angle-side" in common.

Definition 11.1.3. A triangle is said to be *isosceles* if two of its sides have the same length. The angles opposite the equal sides of an isosceles triangle are called the *base angles* of the triangle.

Theorem 11.1.4. *The base angles of an isosceles triangle are equal.*

Proof. Let the given triangle be $\triangle ABC$ with $AB = AC$. Turn the triangle over and denote the corresponding triangle as $\triangle A'C'B'$, as shown in Figure 11.3. Then $\triangle ABC \cong \triangle A'C'B'$, since $\angle A = \angle A'$ and $AB = A'C' = AC = A'B'$. Thus, they have side-angle-side in common. In this congruence, $\angle B$ corresponds to $\angle C'$ and $\angle C$ to $\angle B'$, so $\angle B = \angle C'$ and $\angle C = \angle B'$. On the other hand, $\angle C'$ was obtained by turning $\angle C$ over, and so $\angle C' = \angle C$. It follows that $\angle B = \angle C$, as was to be proven. $\square$

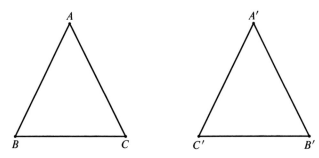

Fig. 11.3 Proving that the base angles of an isosceles triangle are equal

Definition 11.1.5. A triangle is *equilateral* if all three of its sides have the same length.

Corollary 11.1.6. *All three angles of an equilateral triangle are equal to each other.*

Proof. Any two angles of an equilateral triangle are the base angles of an isosceles triangle and are therefore equal to each other by the previous theorem. It follows that all three angles are equal. $\square$

It is sometimes convenient to establish congruence of triangles by correspondences other than side-angle-side.

Theorem 11.1.7 (Angle-Side-Angle). *If two triangles have "angle-side-angle" in common, then they are congruent.*

Proof. Suppose that triangles ABC and DEF are given with $\angle A = \angle D$, $AB = DE$, and $\angle E = \angle B$. If $AC = DF$, then the triangles are congruent by side-angle-side. If this is not the case, then one of them is longer; suppose, without loss of generality, that AC is shorter than DF. We will show that is impossible.

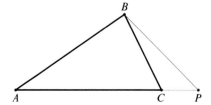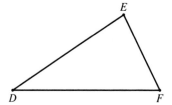

Fig. 11.4 Proving angle-side-angle

Mark the length DF along AC beginning with the point A and ending at a point P, as shown in Figure 11.4. Then draw the line connecting B to P. It would follow that $\triangle ABP$ has side-angle-side in common with $\triangle DEF$, and this would imply that $\angle ABP = \angle E$. But we are assuming that $\angle ABC = \angle E$. This would give $\angle ABC = \angle ABP$, from which we conclude that $\angle PBC = 0$. Therefore, PB lies on BC and hence $AP = AC$.   □

If two triangles have equal sides, then they automatically also have equal angles.

Theorem 11.1.8 (Side-Side-Side). *If two triangles have corresponding sides equal to each other, then they are congruent.*

Proof. Let triangles ABC and DEF be given with $AB = DE$, $BC = EF$, and $AC = DF$. At least one of the sides is greater than or equal to each of the other two; suppose, for example, that AB is greater than or equal to each of AC and CB (the other cases would be proven in exactly the same way). Then place the triangle DEF under $\triangle ABC$ so that DE coincides with AB as in Figure 11.5. Connect the

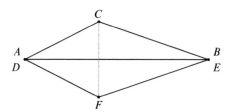

Fig. 11.5 Proving side-side-side

points C and F by a straight line. Since $AC = DF$, triangle AFC is isosceles and the base angles ACF and AFC are equal to each other (by Theorem 11.1.4). Similarly, $\triangle BCF$ is isosceles, so $\angle BCF = \angle BFC$. Adding the angles shows that $\angle ACB = \angle DFE$. It follows that triangles ABC and DEF agree in side-angle-side and are therefore congruent to each other.   □

Definition 11.1.9. A *straight angle* is an angle that is a straight line. That is, the angle ABC is a straight angle if the points A, B, and C all lie on a straight line and B is in between A and C. A *right angle* is an angle that is half the size of a straight angle.

Definition 11.1.10. *Vertical angles* are pairs of angles that occur opposite each other when two lines intersect. In Figure 11.6, the angles *BEA* and *CED* are a pair of vertical angles, and the angles *BED* and *CEA* are a pair of vertical angles.

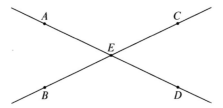

Fig. 11.6 Illustrating vertical angles

Theorem 11.1.11. *Vertical angles are equal.*

Proof. In Figure 11.6, we show that $\angle BEA = \angle CED$, as follows. Angle *BEA* and angle *AEC* add up to a straight angle. Angle *AEC* and angle *CED* also add up to a straight angle. Therefore, $\angle BEA + \angle AEC = \angle AEC + \angle CED$. Hence, angle *BEA* equals angle *CED*.   □

One customary way of denoting the size of angles is in terms of degrees.

Definition 11.1.12. The measure of an angle in *degrees* is defined so that a straight angle is 180° and other angles are the number of degrees determined by the proportion that they are of straight angles. In particular, a right angle is 90°.

We will prove that the sum of the angles of a triangle is a straight angle. In the approach that we follow, the following partial result is essential.

Theorem 11.1.13. *The sum of any two angles of a triangle is less than* 180°.

Proof. Consider an arbitrary triangle *ABC* as depicted in Figure 11.7 (on the next page) and extend the side *AB* beyond *A* as shown. We will show that the sum of angles *CAB* and *ACB* is less than a straight angle.

Let *M* be the midpoint of the side *AC*. Draw the line from *B* through *M* and extend it to the other side of *M* to a point *D* such that $DM = MB$. Draw the line from *D* to *A*. Then $\angle DMA = \angle CMB$, since they are a pair of vertical angles (Theorem 11.1.11). By construction, $AM = MC$ and $DM = MB$. Thus, $\triangle CMB \cong \triangle AMD$ by side-angle-side (11.1.2). It follows that $\angle DAM$ is equal to $\angle BCM$. Therefore, the sum that we are interested in, $\angle BCM + \angle MAB$, is equal to the sum of $\angle DAM + \angle MAB$. But this latter sum is less than a straight angle, since it together with $\angle DAF$ sums to a straight angle.   □

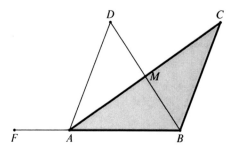

Fig. 11.7 The sum of two angles of a triangle is less than 180 degrees

11.2   The Parallel Postulate

By a *line* we mean a straight line extending infinitely in both directions; by a *line segment*, we mean a finite part of a line between two given points. Two lines in the plane are *parallel* if they do not intersect.

For hundreds of years, mathematicians tried to prove the following as a theorem that followed from the other basic assumptions about Euclidean geometry. Finally, in the 1880s, this was shown to be impossible when different geometries were constructed that satisfied the other basic assumptions but not the following (they are now called "non-Euclidean geometries"). Since it cannot be proven, we assume it as an axiom.

The Parallel Postulate 11.2.1. Given a line and a point that is not on the line, there is one and only one line through the given point that is parallel to the given line.

We will develop necessary and sufficient conditions that two lines be parallel. Given two lines, a third line that intersects both of the first two is said be a *transversal* of the two lines. Given a transversal of two lines, a pair of angles created by the intersections of the transversal with the lines are said to be *corresponding angles* if they lie on the same sides of the given lines. In Figure 11.8, T is a

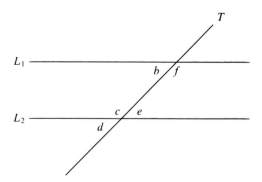

Fig. 11.8 Corresponding angles and alternate interior angles

transversal of the lines L_1 and L_2. The angles b and d are a pair of corresponding angles. The four angles between the lines are called *interior angles*. If two interior angles lie on opposite sides of the transversal, they are called *alternate interior angles*. In Figure 11.8, the angles b and e are a pair of alternate interior angles, as are the angles c and f.

Theorem 11.2.2. *If the angles in a pair of corresponding angles created by a transversal of two lines are equal to each other, then the given lines are parallel.*

Proof. If the theorem was not true, then there would be a situation as depicted in Figure 11.9, where $\angle a = \angle c$ and lines L_1 and L_2 intersect in some point P.

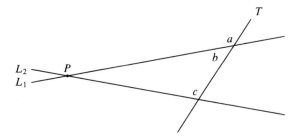

Fig. 11.9 Equal corresponding angles imply that lines are parallel

Now $\angle a + \angle b$ is clearly a straight angle. Then, since $\angle a = \angle c$, it would follow that that the sum of angles b and c is a straight angle, contradicting Theorem 11.1.13. Hence, the lines L_1 and L_2 cannot intersect.   □

The converse of this theorem is also true.

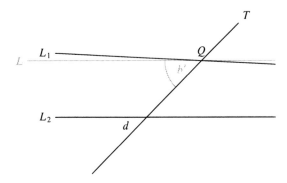

Fig. 11.10 If lines are parallel, corresponding angles are equal

Theorem 11.2.3. *If two lines are parallel, then any pair of corresponding angles are equal to each other.*

Proof. Suppose that two lines are parallel and that two corresponding angles differ from each other. Then there would be a situation, such as that depicted in Figure 11.8, with two parallel lines L_1 and L_2 and angle b different from angle d. Suppose that angle b is bigger than angle d (the proof where this inequality is reversed would be virtually identical). Then, as depicted above in Figure 11.10, we could draw a line L through Q, the intersection point of L_1 and T such that angle b' is equal to angle d.

But then, by the previous theorem (11.2.2), L would be parallel to L_2. Thus, L and L_1 would be distinct lines through the point Q both of which are parallel to L_2, contradicting the uniqueness aspect of the Parallel Postulate (11.2.1).   □

Corollary 11.2.4. *If two lines are parallel, then any pair of alternate interior angles are equal to each other.*

Proof. Consider the alternate interior angles b and e in Figure 11.8. From Theorem 11.2.3, we know that angles b and d are equal, and by Theorem 11.1.11, angle d is equal to angle e. Therefore, angles b and e are equal.   □

We can now establish the fundamental theorem on the angles of a triangle.

Theorem 11.2.5. *The sum of the angles of a triangle is a straight angle.*

Proof. Let a triangle ABC be given. Use the Parallel Postulate (11.2.1) to pass a line through A that is parallel to BC and mark points D and E on that line, as in Figure 11.11. By Corollary 11.2.4, $\angle DAB = \angle ABC$ and $\angle EAC = \angle ACB$. Clearly, the sum of the angles DAB, BAC, and EAC is a straight angle. Hence, the sum of the angles ABC, BAC, and ACB is also a straight angle.   □

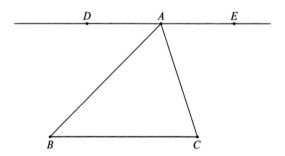

Fig. 11.11 The sum of the angles of a triangle is a straight angle

The following is an obvious corollary.

Corollary 11.2.6. *If two angles of one triangle are respectively equal to two angles of another triangle, then the third angles of the triangles are also equal.*

Corollary 11.2.7 (Angle-Angle-Side). *If two triangles agree in angle-angle-side, then they are congruent.*

Proof. By the previous corollary, the triangles have their third angles equal as well. Thus, the triangles also agree in angle-side-angle, and by Theorem 11.1.7, they are congruent.   □

11.3   Areas of Triangles

We require knowledge of the areas of some common geometric figures. We begin with the following definition which forms the basis for the definition of areas of all geometric figures.

Definition 11.3.1. The *area of a rectangle* is defined to be the product of its length and its width.

The areas of other geometric figures can be obtained either by directly comparing them to rectangles or by approximating them by rectangles.

Definition 11.3.2. Lines, or line segments, are said to be *perpendicular* (or *orthogonal*) if they intersect in a right angle.

Definition 11.3.3. A *right triangle* is a triangle one of whose angles is a right angle. The side opposite the right angle in a right triangle is called the *hypotenuse* of the triangle, and the other two sides are called the *legs*.

Theorem 11.3.4. *The area of a right triangle is one-half the product of the legs of the triangle.*

Proof. Let the right triangle $\triangle ABC$ be as pictured in Figure 11.12. By creating perpendiculars to AC at A and to BC at B, complete the triangle to a rectangle as shown. Since the sum of the angles of a triangle is 180° (Theorem 11.2.5), the sum of angles BAC and ABC is 90°. Since AD is perpendicular to AC, the sum of the

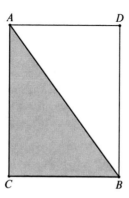

Fig. 11.12 Area of a right triangle

angles BAC and BAD is also 90°. Hence, $\angle ABC = \angle BAD$. Since angle CBD is a right angle, $\angle CAB = \angle ABD$, as they both sum with angle ABC to 90°. It follows that $\triangle ABC \cong \triangle BAD$, since they agree in angle-side-angle (11.1.7). Thus, those triangles have equal areas. Since their areas sum to the area of the rectangle whose area is the product of the legs of the triangle ABC, it follows that the area of the triangle ABC is one-half of that product.   □

Any one of the sides of a triangle may be regarded as a *base* of the triangle.

Definition 11.3.5. If a side of a triangle is designated as its *base*, then the *height* of the triangle (relative to that base) is the length of the perpendicular from the base to the vertex of the triangle not on the base. It may be necessary to extend the base of the triangle in order to determine its height, as in the second triangle pictured in Figure 11.13. (In both of the triangles depicted in Figure 11.13, h is the height of the triangle to the base AC.)

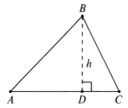 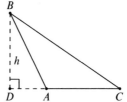

Fig. 11.13 Heights of triangles

Theorem 11.3.6. *The area of any triangle is one-half the product of a base of the triangle and the height of the triangle to that base.*

Proof. Suppose that the triangle ABC is as pictured in the first triangle in Figure 11.13, where h is the height to the base AC. Then, by the previous theorem (11.3.4), the area of the right triangle ABD is one-half the product of h and AD, and the area of right triangle DBC is one-half the product of h and DC. The area of triangle ABC is the sum of those areas and is therefore $\frac{1}{2}h \cdot (AD) + \frac{1}{2}h \cdot (DC) = \frac{1}{2}h \cdot (AD + DC) = \frac{1}{2}h \cdot (AC)$. This finishes the proof in this case.

Suppose that $\triangle ABC$ is as pictured in the second triangle in Figure 11.13. The side AC had to be extended to the point D at the bottom of the height. In this case, the area of $\triangle ABC$ is the difference between the area of the right triangle BDC and the area of the right triangle BDA. Hence, the area is $\frac{1}{2}h \cdot (DC) - \frac{1}{2}h \cdot (DA) = \frac{1}{2}h \cdot (AC)$.   □

One of the most famous theorems in mathematics is the Pythagorean Theorem. The easiest way to prove it is by using areas.

The Pythagorean Theorem 11.3.7. *For any right triangle, the square of the length of the hypotenuse is equal to the sum of the squares of the lengths of the legs.*

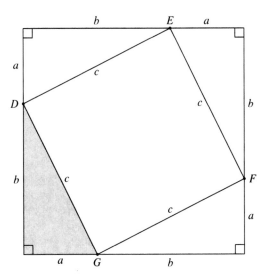

Fig. 11.14 Proof of the Pythagorean Theorem

Proof. Let the right triangle have legs of length a and b and hypotenuse of length c. This proof of the Pythagorean Theorem is obtained by placing four copies of the given right triangle inside a square whose sides have length $a + b$, as shown in Figure 11.14. We need to prove that the four-sided figure $DEFG$ is a square; i.e., since each of its sides has length c, we must prove that each of its angles is a right angle. But this follows immediately from the fact that each such angle sums with the two non-right angles of the original triangle to a straight angle. Thus, $DEFG$ is a square, each of whose sides has length c. The area of the big square, each of whose sides has length $a + b$, is the sum of the area of the square $DEFG$ and four times the area of the original right triangle. That is, $(a + b)^2 = 4(\frac{1}{2}ab) + c^2$. Thus, $a^2 + 2ab + b^2 = 2ab + c^2$, or $a^2 + b^2 = c^2$.   □

Definition 11.3.8. Two triangles are *similar* if their vertices can be paired so that the corresponding angles are equal to each other. We use the notation $\triangle ABC \sim \triangle DEF$ to denote similarity.

Of course (by Corollary 11.2.6), it follows that two triangles are similar if they agree in two of their angles. It is an important, and nontrivial, fact that the corresponding sides of similar triangles are proportional to each other. In other words, if $\triangle ABC \sim \triangle DEF$, then $\frac{AB}{DE} = \frac{AC}{DF} = \frac{BC}{EF}$. The ingenious proof that we present goes back to Euclid.

We begin with a lemma.

Lemma 11.3.9. *If two lines are parallel and two other lines are perpendicular to the parallel lines, then the lengths of the perpendicular line segments between the parallel lines are equal to each other.*

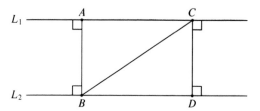

Fig. 11.15 Perpendiculars between parallel lines are equal

Proof. In Figure 11.15, we are assuming that L_1 is parallel to L_2 and that AB and DC are perpendicular to both of L_1 and L_2. (By Theorem 11.2.3, if a line is perpendicular to one of two parallel lines, it is perpendicular to the other as well.) We must prove that $AB = CD$. Note that $\angle ACB = \angle DBC$ and $\angle ABC = \angle BCD$, by Corollary 11.2.4. Thus, the triangles ABC and BCD are congruent by angle-side-angle, since they also share the side BC. Therefore, the corresponding sides AB and CD are equal to each other. □

Our basic approach to the proportionality theorem is based on the following lemma.

Lemma 11.3.10. *If a triangle with area S_1 has the same height with respect to a base b_1 that a triangle with area S_2 has with respect to its base b_2, then $\frac{S_1}{b_1} = \frac{S_2}{b_2}$.*

Proof. Let the common height of the two triangles with respect to the given bases be h. Then, $S_1 = \frac{1}{2}hb_1$ and $S_2 = \frac{1}{2}hb_2$. It follows that $\frac{S_1}{b_1} = \frac{1}{2}h = \frac{S_2}{b_2}$. □

Theorem 11.3.11. *If two triangles are similar, then their corresponding sides are proportional. That is, if $\triangle ABC \sim \triangle DEF$, then $\frac{AB}{DE} = \frac{AC}{DF} = \frac{BC}{EF}$.*

Proof. It suffices to prove that $\frac{AB}{DE} = \frac{AC}{DF}$; the other equation can be obtained as in the proof below but placing the triangles so that the angle at B coincides with the angle at E.

Place the triangles so that the angle of the first triangle at A coincides with the angle of the second triangle at D. If the length of AB is the same as the length of DE, then the two triangles are congruent and all the proportions are 1. Assume, then, that the length of AB is less than the length of DE. (If the opposite is true, the proof below can be accomplished by interchanging the roles of $\triangle ABC$ and $\triangle DEF$.) The situation is depicted in Figure 11.16.

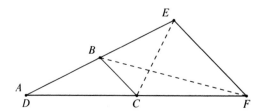

Fig. 11.16 Corresponding sides of similar triangles are proportional

We need to construct triangles to which we can apply the preceding lemma. In Figure 11.16, connect B and F by a line and C and E by a line. Note that by Theorem 11.2.2, $\angle ABC = \angle DEF$ implies that the line BC is parallel to the line EF. Regard the triangles BEC and BFC as having a common base BC. Then the corresponding heights of the triangles are the perpendiculars from E to (the extension of) BC and from F to (the extension of) BC, respectively. By Lemma 11.3.9 those heights are equal to each other. Thus, triangles BEC and BFC, having equal bases and heights, have equal areas. Adding those triangles to $\triangle ABC$ establishes that triangles ACE and ABF have equal areas.

We can now use Lemma 11.3.10, as follows. Since $\triangle ABC$ has the same height with respect to its base AB as $\triangle ACE$ has with respect to its base DE, Lemma 11.3.10 implies that

$$\frac{\text{area}(\triangle ABC)}{AB} = \frac{\text{area}(\triangle ACE)}{DE}, \text{ or } \frac{AB}{DE} = \frac{\text{area}(\triangle ABC)}{\text{area}(\triangle ACE)}$$

Similarly, $\triangle ABC$ has the same height with respect to its base AC as $\triangle ABF$ has with respect to its base DF, so

$$\frac{\text{area}(\triangle ABC)}{AC} = \frac{\text{area}(\triangle ABF)}{DF}, \text{ or } \frac{AC}{DF} = \frac{\text{area}(\triangle ABC)}{\text{area}(\triangle ABF)}$$

Since the triangles ABF and ACE have the same area, it follows that $\frac{AB}{DE} = \frac{AC}{DF}$.
□

We will need the following result in Chapter 12.

Theorem 11.3.12. *If an angle is inscribed in a circle and the arc that it cuts off is a semicircle (see Figure 11.17), then the angle is a right angle.*

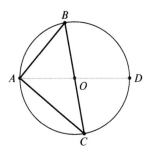

Fig. 11.17 An inscribed right angle

Proof. The angles that we are considering are those angles such as $\angle BAC$ in Figure 11.17, where BC is the diameter of the circle and O is the center of the circle. Draw the diameter from A through O.

Note that $\angle OAB = \angle OBA$, since $OA = OB$, and similarly $\angle OAC = \angle OCA$. Moreover, the angles $\angle AOC$ and $\angle BOD$ are equal to each other, since they are a pair of vertical angles (Theorem 11.1.11), as are the angles $\angle AOB$ and $\angle COD$. Thus, we can label the angles x, y, z, w as indicated in Figure 11.18.

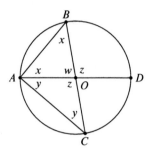

Fig. 11.18 Proving an inscribed angle is a right angle

Since $2x + w = 180°$ and $z + w = 180°$, it follows that $z = 2x$. Similarly, $w = 2y$. Therefore, $2x + 2y = z + w = 180°$, so $x + y = 90°$. This shows that $\angle BAC$ is $90°$.   □

11.4 Problems

Basic Exercises

1. Which of the following triples cannot be the lengths of the sides of a right triangle?

   (a) $3, 4, 5$
   (b) $1, 1, 1$
   (c) $2, 3, 4$
   (d) $1, \sqrt{3}, 2$

2. In the diagram given below, lines L_1 and L_2 are parallel and line T is a transversal. If the measure of $\angle d$ is $55°$ and the measure of $\angle f$ is $130°$, find the measures of $\angle b$, $\angle e$, and $\angle c$.

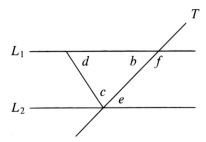

3. In the diagram given below, the line segment BD is perpendicular to the line segment AC, the length of AM is equal to the length of MC, the measure of $\angle C$ is 35°, and the measure of $\angle FAD$ is 111°.

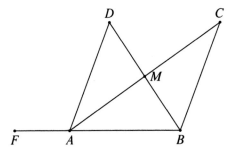

   (a) Prove that triangle ABM is congruent to triangle CBM.
   (b) Find the measure of $\angle CAB$.
   (c) Find the measure of $\angle ABC$.
   (d) Find the measure of $\angle ABD$.
   (e) Find the measure of $\angle AMD$.
   (f) Find the measure of $\angle D$.
   (g) Show that the line segments AD and BC are not parallel.

4. Prove that two right triangles are congruent if they have equal hypotenuses and a pair of equal legs.

Interesting Problems

5. A *quadrilateral* is a four-sided figure in the plane. Prove that the sum of the angles of a quadrilateral is 360°.
6. For quadrilateral $ABCD$, as shown below, suppose that $\angle ABC = \angle CDA$ and $\angle DAB = \angle BCD$. Prove that $AB = CD$ and $BC = AD$.

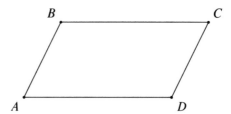

7. Prove that if two angles of a triangle are equal, then the sides opposite those angles are equal.
8. With reference to the diagram below, prove that $\triangle ABC$ is an isosceles triangle if $\angle DAB = \angle EAC$ and DE is parallel to BC.

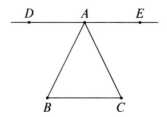

9. A *parallelogram* is a four-sided figure in the plane whose opposite sides are parallel to each other. Prove the following:

   (a) The opposite sides of a parallelogram have the same length.
   (b) The area of a parallelogram is the product of the length of any side and the length of a perpendicular to that side from a vertex not on that side.
   (c) If one of the angles of a parallelogram is a right angle, then the parallelogram is a rectangle.

10. A *trapezoid* is a four-sided figure in the plane two of whose sides are parallel to each other. The *height* of a trapezoid is the length of a perpendicular from one of the parallel sides to the other. Prove that the area of a trapezoid is its height multiplied by the average of the lengths of the two parallel sides.
11. A *square* is a four-sided figure in the plane all of whose sides are equal to each other and all of whose angles are right angles. The *diagonals* of the square are the lines joining opposite vertices. Prove that the diagonals of a square are perpendicular to each other.
12. Show that lines are parallel if there is a transversal such that the alternate interior angles are equal to each other.

Challenging Problems

13. Give an example of two triangles that agree in "angle-side-side" but are not congruent to each other.

14. Find the length of line segment OA in the diagram below (line segment OD is a radius of the circle centered at O, line segment AD has length 4, line segment AC has length 10, and $OABC$ is a rectangle).

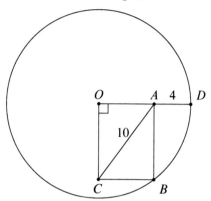

15. Prove the converse of the Pythagorean Theorem; i.e., show that if the lengths of the sides of a triangle satisfy the equation $a^2 + b^2 = c^2$, then the triangle is a right triangle.

16. The following problem provides some basic results in trigonometry

   (a) Let θ be any angle between 0 and 90°. Place θ in a right triangle, as shown in the diagram below, and label the sides as in the diagram. Define $\sin \theta$ to be $\frac{a}{c}$, $\cos \theta$ to be $\frac{b}{c}$, and $\tan \theta$ to be $\frac{a}{b}$. Using Theorem 11.3.11, show that these definitions do not depend on which right triangle θ is placed in.

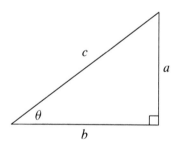

   (b) Label the angles of a triangle with A, B, and C and label the side opposite $\angle A$ with a, the side opposite $\angle B$ with b, and the side opposite $\angle C$ with c. Prove that, in the case where the angles A, B and C are all less than 90 degrees, (although the results are true for all triangles):

   (i) $\dfrac{a}{\sin A} = \dfrac{b}{\sin B} = \dfrac{c}{\sin C}$ (The Law of Sines)

   (ii) $c^2 = a^2 + b^2 - 2ab \cos C$ (The Law of Cosines)

17. (This problem generalizes the result of Theorem 11.3.12.) Prove that the measure of an angle inscribed in a circle is one-half the measure of the arc

cut off by the angle. That is, in the diagram below, the number of degrees of
$\angle BAC$ is half the number of degrees in the arc BC. (The number of degrees
in a full circle is 360, and the number of degrees in any arc of a circle is the
product of 360 and the length of that arc divided by the circumference of the
circle.)

[Hint: One approach is by first proving the special case where AC is a diameter
of the circle.]

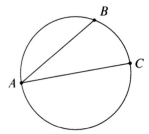

Chapter 12
Constructability

The Ancient Greeks were interested in many different kinds of mathematical problems. One of the aspects of geometry that they investigated is the question of what geometric figures can be constructed using a compass and a straightedge. A *compass* is an instrument for drawing circles. The compass has two branches that open up like a scissors. One of the branches has a sharp point at the end and the other branch has a pen or pencil at the end. If the compass is opened so that the distance between the two ends is r and the pointed end is placed on a piece of paper and the compass is twirled about that point, the writing end traces out a circle of radius r. The drawing made by any real compass will only approximate a circle of radius r. But we are going to consider constructions theoretically; we will assume that a compass opened up a distance r precisely makes a circle of radius r.

To do geometric constructions, we will also require (as the Ancient Greeks did) another implement. By a *straightedge* we will mean a device for drawing lines connecting two points and extending such lines as far as desired in either direction. Sometimes people inaccurately speak of constructions with "ruler and compass." It is important to understand that the constructions investigated by the Ancient Greeks do not allow use of a ruler in the sense of an instrument that has distances marked on it. We can only use such an instrument to connect pairs of points by straight lines; we cannot use it to measure distances.

In this chapter, when we say "construct" or "construction," we always mean "using only a compass and a straightedge."

We will indicate how to do some basic constructions. But the most interesting part of this chapter will be proving that certain geometric objects cannot be constructed. In particular, we will prove that an angle of 20° cannot be constructed. This implies that an angle of 60° cannot be trisected (i.e., divided into three equal parts) with a straightedge and compass. The Ancient Greeks assumed that there must be some way of trisecting every angle; they thought that they had simply not been clever enough to find a method for doing so. It was only after mathematical advances in the nineteenth century that it could be proven that there is no way to trisect an angle of 60° with a straightedge and compass. The highlight of this chapter will be

D. Rosenthal et al., *A Readable Introduction to Real Mathematics*,
Undergraduate Texts in Mathematics, DOI 10.1007/978-3-319-05654-8_12,
© Springer International Publishing Switzerland 2014

a proof of that fact. Although it is hard to imagine how something like that could be proved, we shall see that there is an indirect approach that also establishes many other interesting results.

12.1 Constructions with Straightedge and Compass

Let's start with some very basic constructions.

Definition 12.1.1. A *perpendicular bisector* of a line segment is a line that is perpendicular to the line segment and goes through the middle of the line segment.

Theorem 12.1.2. *Given any line segment, its perpendicular bisector can be constructed.*

Proof. Given a line segment AB, as shown in Figure 12.1, put the point of the compass at A and open the compass to radius the length of AB. Let r equal the length of AB. Then draw the circle with center at A and radius r. Similarly, draw the circle with center at B and radius r. The two circles will intersect at points, C and D, as indicated in Figure 12.1. Take the straightedge and draw the line segment from C to D. We claim that CD is a perpendicular bisector of AB.

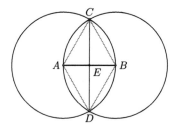

Fig. 12.1 Constructing the perpendicular bisector of a line segment

To prove this, label the point of intersection of CD and AB as E and then draw the line segments AC, CB, BD, and DA. We must prove that $AE = EB$ and that $\angle CEA$ (and/or any of the other three angles at E) is a right angle. First note that AC, CB, BD, and DA all have the same length, r, since they are all radii of the two circles of radius r. Thus, triangle ACD is congruent to triangle DCB, since the third side of each is CD and they therefore agree by side-side-side (11.1.8). It follows that $\angle ACE = \angle BCE$. Thus, triangle ACE is congruent to triangle BCE by side-angle-side (11.1.2). Therefore, $AE = EB$. Moreover, $\angle AEC = \angle BEC$, so, since those two angles sum to a straight angle, each of them is a right angle.   □

Definition 12.1.3. An *angle bisector* is a line from the vertex of the angle that divides it into two equal subangles.

Theorem 12.1.4. *Given any angle, its bisector can be constructed.*

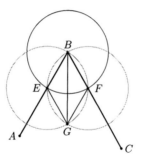

Fig. 12.2 Constructing the bisector of an angle

Proof. Consider an angle ABC, as pictured in Figure 12.2, and draw a circle centered at B that intersects both BA and BC. Label the points of intersection of the circle with AB and with BC as E and F, respectively. Let r be the distance from E to F. Use the compass to draw a circle of radius r centered at E and a circle of radius r centered at F. These two circles intersect in some point G within the angle ABC, as shown in Figure 12.2. Use the straightedge to draw the line segment connecting B to G. We claim that this line segment bisects the angle ABC.

Draw the lines EG and FG. We prove that triangle BEG is congruent to triangle BFG. Note that $BE = BF$, since they are both radii of the original circle centered at B. Note also that $EG = FG$, since they are each radii of circles with radius r. Since triangle BEG and triangle BFG share side BG, it follows from side-side-side (11.1.8) that the two triangles are congruent. Thus, $\angle EBG = \angle FBG$ and BG is a bisector of angle ABC. □

Theorem 12.1.5. *Any given line segment can be copied using only a straightedge and compass.*

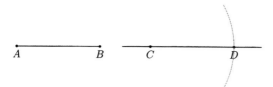

Fig. 12.3 Copying a line segment

Proof. Suppose a line segment AB is given, as pictured in Figure 12.3, and it is desired to copy it on another line. Choose any point C on the other line, and then open the compass to a radius the length of AB. Put the point of the compass at C and draw any portion of the resulting circle that intersects the other line. Label the point of intersection D. Then CD is copy of AB. □

Theorem 12.1.6. *Any given angle can be copied using only a straightedge and compass.*

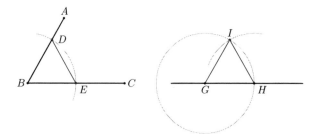

Fig. 12.4 Copying an angle

Proof. Let an angle ABC be given, as in Figure 12.4. We construct an angle equal to $\angle ABC$ with vertex G on any other line. To do this, draw any arc of any circle (of radius, say, r) centered at B that intersects both BA and BC. Label the points of intersection D and E. Draw the circle of radius r centered at G. Use H to label the point where that circle intersects the line containing G. Then adjust the compass to be able to make circles of radius DE. Put the point of the compass at H and draw a portion of the circle that intersects the circle centered at G; call that point of intersection I. Draw line segments connecting D to E and I to H.

Then $IH = DE$, since IH is a radius of a circle with radius DE. Also draw the line segment GI. The lengths of BD, BE, GI, and GH are all equal to r. It follows by side-side-side (11.1.8) that triangle BDE is congruent to triangle GIH. Thus, $\angle IGH$ is a copy of $\angle ABC$.   □

Corollary 12.1.7. *If the angles α and β are constructed, then:*

(i) the angle $\alpha + \beta$ can be constructed, and
(ii) for every natural number n, the angle $n\alpha$ can be constructed.

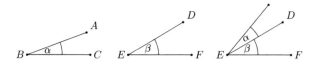

Fig. 12.5 Constructing the sum of two angles

Proof. (i) Let the angles α and β be given, as pictured in Figure 12.5. To construct the angle $\alpha + \beta$, simply copy the angle α with one side DE and the other side outside the original angle β, as shown in the third diagram in Figure 12.5.

(ii) This clearly follows from repeated application of part (i), starting with angles α and β that are equal to each other. (This can be proven more formally using mathematical induction.)   □

Theorem 12.1.8. *Given any line segment and any natural number n, the line segment can be divided into n equal parts using only a straightedge and compass.*

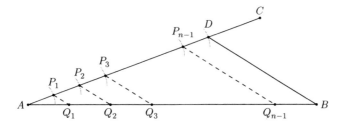

Fig. 12.6 Dividing a line segment into n equal parts

Proof. Fix a natural number n. Let a line segment AB be given, as shown in Figure 12.6. Use the straightedge to draw any line segment emanating from A that is at a positive angle with AB, and pick a point C on it as shown. Open the compass to any radius s less than one n^{th} of the length of AC. Beginning at A use the compass to mark off n consecutive segments of AC of length s, as illustrated in Figure 12.6. Label the points of intersection of the arcs and AC as $P_1, P_2, P_3, \ldots, P_{n-1}$. Label with D the point of intersection of the line and the last arc drawn. Use a straightedge to connect D to B. We then construct lines parallel to DB through each point of intersection of an arc with AD, after which we will show that the intersections of those lines with AB divide AB into n equal segments.

To construct the parallel lines, copy the angle ADB (Theorem 12.1.6) at each point of intersection of an arc with AD so that one side of the new angle lies on AD and the other side points downwards and is extended to intersect the line AB. These are the dotted lines in Figure 12.6. Label the points of intersection of the dotted lines with AB as $Q_1, Q_2, Q_3, \ldots, Q_{n-1}$, as shown. We claim that the points $\{Q_1, Q_2, Q_3, \ldots, Q_{n-1}\}$ divide the segment AB into n equal parts. To see this, note that for each j, the triangle AP_jQ_j has two angles, $\angle P_jAQ_j$ and $\angle AP_jQ_j$, equal to corresponding angles of $\triangle ADB$. Thus, $\triangle AP_jQ_j$ is similar to $\triangle ADB$ (Corollary 11.2.6). Therefore, the corresponding sides are proportional (Theorem 11.3.11). The ratio of AP_j to AD is $\frac{j}{n}$. Thus, the length of AQ_j divided by the length of AB is also $\frac{j}{n}$.   □

Even though a line segment can be divided into any number of equal parts, some angles, such as those of 60 degrees, cannot be divided into three equal parts using only a straightedge and compass. We now begin preparation for an indirect approach to establishing that fact.

12.2   Constructible Numbers

We consider constructing numbers instead of constructing geometric objects, although we will use geometric constructions to construct the numbers.

We begin by imagining a horizontal line on which a point is arbitrarily marked as 0 and another point, to the right of it, is arbitrarily marked as 1. We consider the question of what other numbers can be obtained by starting with the length 1 (that we take as the distance between the points marked 0 and 1) and doing *geometric constructions* in the plane to obtain other lengths. By a geometric construction, we mean using our straightedge to make lines joining any two points we have already marked (i.e., constructed) or using our compass to construct a circle centered at a constructed point using a radius that has been constructed.

Definition 12.2.1. A real number is *constructible* if the point corresponding to it on the number line can be obtained from the marked points 0 and 1 by performing a finite sequence of constructions using only a straightedge and compass.

Theorem 12.2.2. *Every integer is constructible.*

Proof. The numbers 0 and 1 are given as constructible. The number 2 can easily be constructed: simply take a compass, open it up to radius 1 by placing one side at the point 0 and the other side at the point 1, and then place the pointed side on the point marked 1 and draw the circle of radius 1 with that point as center. The point where that circle meets the number line to the right of 1 is the number 2, so 2 has been constructed. Then clearly 3 can be constructed by placing the compass with radius 1 so as to make a circle centered at 2. Similarly, all the natural numbers can be constructed. To construct the number -1, simply make the circle of radius 1 centered at 0 and mark the intersection to the left of 0 of that circle with the number line. Then -2 can be constructed by marking the point where the circle centered at -1 meets the number line to the left of the point -1. Every negative integer can be constructed similarly.   □

What about the rational numbers?

Theorem 12.2.3. *Every rational number is constructible.*

Proof. To construct, for example, the number $\frac{1}{3}$, simply divide the interval between 0 and 1 into three equal parts (see Theorem 12.1.8) and mark the right-most point of the first part as $\frac{1}{3}$. Similarly, for any natural number n, dividing the unit interval into n equal parts shows that $\frac{1}{n}$ is constructible. Then, for any natural number m, $\frac{m}{n}$ can be constructed by placing m segments of length $\frac{1}{n}$ next to each other on the number line with the first of those segments beginning at 0.

We have therefore shown that all of the positive rational numbers are constructible. If x is a negative rational number, construct $|x|$ and then make a circle of radius $|x|$ centered at 0; the point to the left of 0 where that circle intersects the number line is x. Thus, every rational number is constructible.   □

We need to get information about the set of all constructible numbers. It is essential to the development of this approach that doing arithmetic with constructible numbers produces constructible numbers.

Theorem 12.2.4. *If a is constructible, then $-a$ is constructible.*

Proof. Place a compass on the number line with its point at 0 and the other end opened to a. Then draw the circle. The number $-a$ will be the point of intersection of the circle and the number line opposite to that of a. (If a is positive, then $-a$ is negative, but if a is negative, then $-a$ is positive.) $\qquad\square$

Theorem 12.2.5. *The sum of two constructible numbers is constructible.*

Proof. Suppose that a and b are constructible. If $b = 0$, then clearly $a + b = a + 0 = a$ is constructible. So assume that $b \neq 0$. Open the compass to radius $|b|$, place the point of the compass on the number line at a, and draw the circle. If b is positive, then $a + b$ will be the point of intersection of the circle and the number line to the right of a. If b is negative, then $a + b$ will be the point of intersection of the circle and the number line to the left of a. In both cases, this proves that $a + b$ is constructible. $\qquad\square$

We also need to construct products and quotients. These constructions are a little more complicated; we begin with the following.

Theorem 12.2.6. *If a and b are positive constructible numbers, then $\frac{a}{b}$ is constructible.*

Proof. We consider the two possible cases, where b is greater than 1 and b is less than 1, respectively.

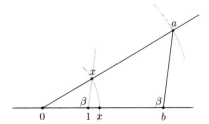

Fig. 12.7 Constructing quotients (first case)

For the case where b is greater than 1, draw the numbers 0, 1, and b on the number line. Use the straightedge to draw a line segment of length greater than a starting from 0, making any angle greater than $0°$ and less than $90°$ with the number line, as pictured in Figure 12.7. Since a is constructible, we can open the compass to radius a. Place the point of the compass at 0 and mark a on the line above the number line. Use the straightedge to connect the point a on the new line to the point b on the number line. Copy the angle, β, at b on the number line to the point 1 on the number line so that the lower side of the angle is the number line itself. Use the straightedge to extend the other side of the angle beyond the new line. The intersection of the other side of the angle and the new line is a point that we have thereby constructed. Let the distance from the origin to that point be x. We can open the compass to radius x and thereby mark x on the number line. So x is a constructible number. The relationship between x and a and b can be determined by observing that the two triangles formed by the above construction are similar to each other, and therefore the corresponding sides are in proportion (Theorem 11.3.11). It follows that $\frac{x}{a} = \frac{1}{b}$. Thus, $x = \frac{a}{b}$, so we have constructed $\frac{a}{b}$.

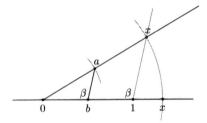

Fig. 12.8 Constructing quotients (second case)

The case where b is less than 1 is very similar. In this case, 1 is to the right of b on the number line. Use the straightedge to make a side of an angle starting at 0 above the number line. Since a is constructible, we can open the compass to radius a and mark a point on the new line that is distance a from the vertex of the angle, as in Figure 12.8. Then use the straightedge to draw a straight line between that point and the point b on the number line. Copy the angle, β, at the point b on the number line to the point 1 on the number line and extend the side of the angle so that it intersects the other line. The compass can then be opened to radius equal to the distance from that point of intersection to the origin. If x denotes that radius, then the fact that the corresponding sides of similar triangles are proportional gives $\frac{a}{x} = \frac{b}{1}$, so that $x = \frac{a}{b}$. Thus, $\frac{a}{b}$ is constructible.   □

It is easy to extend the above to negative numbers.

Corollary 12.2.7. *If a and b are constructible numbers, then ab is constructible and, if $b \neq 0$, $\frac{a}{b}$ is constructible.*

Proof. First suppose that a and b are both positive. Then $\frac{a}{b}$ is constructible by the previous theorem. Let $c = \frac{1}{b}$; then c is constructible by the previous theorem using $a = 1$. Since c is constructible, the previous theorem implies that $\frac{a}{c}$ is constructible. But $\frac{a}{c} = \frac{a}{\left(\frac{1}{b}\right)} = ab$, so ab is constructible.

If one or both of a and b is negative, the above can be applied to $|a|$ and $|b|$. Then $ab = |a| \cdot |b|$ if a and b are both negative, and $ab = -|a| \cdot |b|$ if exactly one of them is negative. Similarly, $\frac{a}{b}$ is equal to one of $\frac{|a|}{|b|}$ or $-\frac{|a|}{|b|}$. Since we can construct the negative of any constructible number (Theorem 12.2.4), it follows that ab and $\frac{a}{b}$ are constructible in this case as well.   □

A "field" is an abstract mathematical concept. In this book we do not need to consider general fields; we only need to consider subfields of $\mathbb{R}$. The following definition forms the basis for the rest of this chapter.

Definition 12.2.8. A *subfield of* $\mathbb{R}$ is a set $\mathcal{F}$ of real numbers satisfying the following properties:

(i) The numbers 0 and 1 are both in $\mathcal{F}$.
(ii) If x and y are in $\mathcal{F}$, then $x + y$ and xy are in $\mathcal{F}$ (i.e., $\mathcal{F}$ is "closed under addition" and "closed under multiplication").

(iii) If x is in $\mathcal{F}$, then $-x$ is in $\mathcal{F}$.

(iv) If x is in $\mathcal{F}$ and $x \neq 0$, then $\frac{1}{x}$ is in $\mathcal{F}$.

In this chapter, we use the word *field* to mean "subfield of $\mathbb{R}$." There are many different subfields of $\mathbb{R}$. Of course, $\mathbb{R}$ itself is a subfield of $\mathbb{R}$. So is the set $\mathbb{Q}$ of rational numbers. It is clear that $\mathbb{R}$ is the biggest subfield of $\mathbb{R}$; it is almost as obvious that $\mathbb{Q}$ is the smallest, in the following sense.

Theorem 12.2.9. *If $\mathcal{F}$ is any subfield of $\mathbb{R}$, then $\mathcal{F}$ contains all rational numbers.*

Proof. To see this, first note that 0 and 1 are in $\mathcal{F}$ and property (ii) of a subfield of $\mathbb{R}$ implies that 2 is in $\mathcal{F}$, and 3 is in $\mathcal{F}$, and so on. That is, $\mathcal{F}$ contains all the natural numbers (this can be formally established by a very easy mathematical induction). Property (iii) then implies that $\mathcal{F}$ contains all integers. By property (iv), $\mathcal{F}$ contains the reciprocals of every integer other than 0, so by property (ii), $\mathcal{F}$ contains all rational numbers.   □

The following is an important fact.

Theorem 12.2.10. *The set of constructible numbers is a subfield of $\mathbb{R}$.*

Proof. This follows immediately from Theorems 12.2.4 and 12.2.5 and Corollary 12.2.7.   □

One of the fundamental theorems in this chapter (Theorem 12.3.12) will provide an alternative characterization of the field of constructible numbers.

Example 12.2.11. The set $\mathbb{Q}(\sqrt{2})$ defined by

$$\mathbb{Q}(\sqrt{2}) = \left\{ a + b\sqrt{2} : a, b \in \mathbb{Q} \right\}$$

is a subfield of $\mathbb{R}$.

Proof. It is clear that $\mathbb{Q}(\sqrt{2})$ contains 0 (since it equals $0 + 0 \cdot \sqrt{2}$) and 1 (since it equals $1 + 0 \cdot \sqrt{2}$). Moreover,

$$(a_1 + b_1\sqrt{2}) + (a_2 + b_2\sqrt{2}) = (a_1 + a_2) + (b_1 + b_2)\sqrt{2}$$

Hence, $\mathbb{Q}(\sqrt{2})$ is closed under addition. Also,

$$(a_1 + b_1\sqrt{2})(a_2 + b_2\sqrt{2}) = (a_1a_2 + 2b_1b_2) + (a_1b_2 + a_2b_1)\sqrt{2}$$

so $\mathbb{Q}(\sqrt{2})$ is closed under multiplication. Furthermore, $-(a + b\sqrt{2}) = (-a) + (-b)\sqrt{2}$.

It remains to be shown that $\frac{1}{a+b\sqrt{2}}$ is in $\mathbb{Q}(\sqrt{2})$, whenever a and b are not both 0. But,

$$\frac{1}{a + b\sqrt{2}} = \frac{a - b\sqrt{2}}{(a + b\sqrt{2})(a - b\sqrt{2})} = \frac{a - b\sqrt{2}}{a^2 - 2b^2} = \frac{a}{a^2 - 2b^2} + \frac{-b}{a^2 - 2b^2}\sqrt{2}$$

which is the sum of a rational number and a number that is the product of rational number and $\sqrt{2}$ and is therefore in $\mathbb{Q}(\sqrt{2})$. Of course, the above expression would not make sense if $a^2 - 2b^2 = 0$. However, this cannot be the case, since $a^2 - 2b^2 = 0$ would imply $\left(\frac{a}{b}\right)^2 = 2$, and we know that $\sqrt{2}$ is irrational (Theorem 8.2.5). $\square$

The field $\mathbb{Q}(\sqrt{2})$ is the field obtained by starting with the field $\mathbb{Q}$ and "adjoining $\sqrt{2}$" to $\mathbb{Q}$; it is called "the extension of $\mathbb{Q}$ by $\sqrt{2}$." This is a special case of a much more general situation.

Theorem 12.2.12. *Let $\mathcal{F}$ be any subfield of $\mathbb{R}$ and suppose that r is a positive number in $\mathcal{F}$. If $\sqrt{r}$ is not in $\mathcal{F}$ and*

$$\mathcal{F}(\sqrt{r}) = \{a + b\sqrt{r} : a, b \in \mathcal{F}\}$$

then $\mathcal{F}(\sqrt{r})$ is a subfield of $\mathbb{R}$.

Proof. The proof is very similar to the proof given above for the special case of $\mathbb{Q}(\sqrt{2})$. It is very easily seen that 0 and 1 are in $\mathcal{F}(\sqrt{r})$ and that $\mathcal{F}(\sqrt{r})$ is closed under addition. To see that it is closed under multiplication, note that

$$(a_1 + b_1\sqrt{r})(a_2 + b_2\sqrt{r}) = (a_1 a_2 + r b_1 b_2) + (a_1 b_2 + a_2 b_1)\sqrt{r}$$

This is in $\mathcal{F}(\sqrt{r})$ since r is in $\mathcal{F}$ and $\mathcal{F}$ itself is a field. Also,

$$\frac{1}{a + b\sqrt{r}} = \frac{a - b\sqrt{r}}{(a + b\sqrt{r})(a - b\sqrt{r})} = \frac{a - b\sqrt{r}}{a^2 - rb^2} = \frac{a}{a^2 - rb^2} + \frac{-b}{a^2 - rb^2}\sqrt{r}$$

Note that $a^2 - rb^2 \neq 0$ unless a and b are both 0, because $\sqrt{r} \notin \mathcal{F}$. (If $a^2 - rb^2 = 0$ and $b \neq 0$, then $\left(\frac{a}{b}\right)^2 = r$, and it would follow that $\sqrt{r} \in \mathcal{F}$.) $\square$

Definition 12.2.13. If $\mathcal{F}$ is a subfield of $\mathbb{R}$ and r is a positive number that is in $\mathcal{F}$ such that $\sqrt{r}$ is not in $\mathcal{F}$, then the field

$$\mathcal{F}(\sqrt{r}) = \{a + b\sqrt{r} : a, b \in \mathcal{F}\}$$

is the field *obtained by adjoining $\sqrt{r}$ to $\mathcal{F}$* and is called *the extension of $\mathcal{F}$ by $\sqrt{r}$.*

Example 12.2.14. Since $\sqrt{5}$ is not an element of $\mathbb{Q}(\sqrt{2})$, the extension of $\mathbb{Q}(\sqrt{2})$ by $\sqrt{5}$ is

$$\{a + b\sqrt{5} : a, b \in \mathbb{Q}(\sqrt{2})\} = \{(c + d\sqrt{2}) + (e + f\sqrt{2})\sqrt{5} : c, d, e, f \in \mathbb{Q}\}$$

For present purposes, we are interested in adjoining square roots to fields of real numbers because that can be done in a "constructible" way.

Theorem 12.2.15. *If r is a positive constructible number, then $\sqrt{r}$ is constructible.*

Proof. Mark the number $1 + r$ on the number line; label it A as in Figure 12.9. Let $M = \frac{r+1}{2}$; M is constructible. Make a circle with center M and radius M. The circle then goes through the point A and also the point corresponding to 0, which

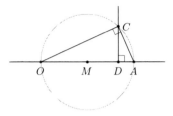

Fig. 12.9 Constructing square roots

we label O. Use D to denote the point corresponding to r on the number line. Erect a perpendicular to the number line at D and let C be the point above the number line at which that perpendicular intersects the circle.

The angle OCA is $90°$, since it is inscribed in a semicircle (Theorem 11.3.12). Therefore, the sum of the angles OCD and DCA is $90°$, from which it follows that the angle COD equals the angle DCA. Thus, triangle OCD is similar to triangle DCA, so their corresponding sides are proportional (Theorem 11.3.11). Let x denote the length of the perpendicular from C to D. Then $\frac{x}{1} = \frac{r}{x}$, so $x^2 = r$. Hence, $x = \sqrt{r}$ and $\sqrt{r}$ is constructible.   □

It follows immediately from this theorem (12.2.15) and the fact that the constructible numbers form a field (Theorem 12.2.10) that every number in $\mathbb{Q}(\sqrt{2})$ is constructible. More generally, every element of $\mathbb{Q}(\sqrt{r})$ is constructible, for every positive rational number r such that $\sqrt{r}$ is irrational. Even more generally, if $\mathcal{F}$ is a field consisting of constructible numbers and r is a positive number in $\mathcal{F}$ such that $\sqrt{r}$ is not in $\mathcal{F}$, then $\mathcal{F}(\sqrt{r})$ consists of constructible numbers. Thus, if we start with $\mathbb{Q}$ and keep on adjoining square roots, we get constructible numbers.

Definition 12.2.16. A *tower of fields* is a finite sequence $\mathcal{F}_0, \mathcal{F}_1, \mathcal{F}_2, \ldots, \mathcal{F}_n$ of subfields of $\mathbb{R}$ such that $\mathcal{F}_0 = \mathbb{Q}$ and, for each i from 1 to n, there is a positive number r_i in $\mathcal{F}_{i-1}$ such that $\sqrt{r_i}$ is not in $\mathcal{F}_{i-1}$ and $\mathcal{F}_i = \mathcal{F}_{i-1}(\sqrt{r_i})$.

Note that a tower can be described as a sequence $\{\mathcal{F}_i\}$ of fields of real numbers such that

$$\mathcal{F}_0 \subset \mathcal{F}_1 \subset \mathcal{F}_2 \subset \cdots \subset \mathcal{F}_n$$

with $\mathcal{F}_0 = \mathbb{Q}$ and each $\mathcal{F}_i$ obtained from its predecessor $\mathcal{F}_{i-1}$ by adjoining a square root.

12.3   Surds

We will show that the constructible numbers are exactly those real numbers that are in fields that are in towers. There is another name that is frequently used for such numbers.

Definition 12.3.1. A *surd* is a number that is in some field that is in a tower. That is, x is a surd if there exists a tower:

$$\mathcal{F}_0 \subset \mathcal{F}_1 \subset \mathcal{F}_2 \subset \cdots \subset \mathcal{F}_n$$

such that x is in $\mathcal{F}_n$.

Theorem 12.3.2. *The set of all surds is a subfield of* $\mathbb{R}$*. Moreover, if r is a positive surd, then* $\sqrt{r}$ *is a surd.*

Proof. To show that the set of surds is a field, it must be shown that the arithmetic operations applied to surds produce surds. This follows immediately if it is shown that for any surds x and y there exists a field $\mathcal{F}$ containing both x and y that occurs in some tower. If $\{\sqrt{r_1}, \sqrt{r_2}, \ldots, \sqrt{r_m}\}$ are the numbers adjoined in making a tower that contains x and $\{\sqrt{s_1}, \sqrt{s_2}, \ldots, \sqrt{s_n}\}$ are the numbers adjoined in making a tower containing y, then adjoining all of those numbers produces a field that contains both x and y. Thus, the set of surds is a subfield of $\mathbb{R}$.

To show that square roots of positive surds are surds, let r be a positive surd. Then r is in some field $\mathcal{F}$ that is in a tower. If $\sqrt{r}$ is in $\mathcal{F}$, then $\sqrt{r}$ is clearly a surd. If $\sqrt{r}$ is not in $\mathcal{F}$, then $\sqrt{r}$ is in $\mathcal{F}(\sqrt{r})$, which is clearly in a tower that has one more field than the tower leading to $\mathcal{F}$. □

Theorem 12.3.3. *Every surd is constructible.*

Proof. This follows immediately from the results that the rational numbers are constructible (Theorem 12.2.3), that the constructible numbers form a field (Theorem 12.2.10), and that the square root of a positive constructible number is constructible (Theorem 12.2.15). □

The fundamental theorem that we will need is that the constructible numbers are exactly the surds. To establish this, we must show that starting with the numbers 0 and 1 and performing constructions with straightedge and compass never produces any numbers that are not surds. Since constructions take place in the plane, we will have to investigate what points in the plane can be constructed.

Definition 12.3.4. We say that the point (x, y) in the plane is *constructible* if that point can be obtained from the points $(0,0)$ and $(1,0)$ by performing a finite sequence of constructions with straightedge and compass.

Theorem 12.3.5. *The point* (x, y) *is constructible if and only if both of the coordinates* x *and* y *are constructible numbers.*

Proof. If x and y are constructible numbers, then the point (x, y) can be constructed by constructing the point x on the x-axis, erecting a perpendicular to the x-axis at the point x and constructing y on that perpendicular.

Conversely, if the point (x, y) has been constructed, then the number x can be constructed by dropping a perpendicular from (x, y) to the x-axis and the number y can be constructed by dropping a perpendicular to the y-axis. (To drop a perpendicular, make a circle with center at the point (x, y) whose radius is large enough that it intersects the axis at two points. Draw the lines from each of the points to the point (x, y). Bisect the angle formed by the two lines just constructed. The resulting triangles, one on either side of the angle bisector, are congruent to each other since they agree in side-angle-side. This implies that the two angles the angle bisector makes with the axis are equal to each other, and, since they sum to a straight angle, they are therefore each 90 degrees. Hence, the angle bisector is a perpendicular from the point (x, y) to the axis.) $\square$

Definition 12.3.6. The *surd plane* is the set of all points (x, y) in the xy-plane such that the coordinates, x and y, are both surds.

By what we have shown above, every point in the surd plane is constructible. We need to show that every constructible point is in the surd plane.

After we have constructed some points, how can we construct others? We can use our straightedge to make lines joining any two points we have constructed, and we can use our compass to construct a circle centered at a constructible point with a radius that is constructible. New constructible points can then be obtained as points of intersection of lines or circles that we have constructed.

Any one line in the plane has many different equations, as does any one circle. We need to know that there are equations with surd coefficients for all of the lines and circles that arise in constructions.

Theorem 12.3.7. *If a line goes through two points in the surd plane, then there is an equation for that line with surd coefficients.*

Proof. Suppose that (x_1, y_1) and (x_2, y_2) are distinct points in the surd plane. We consider two cases. If $x_1 \neq x_2$, then

$$y - y_1 = \frac{y_2 - y_1}{x_2 - x_1}(x - x_1)$$

is an equation of the line through the points (x_1, y_1) and (x_2, y_2). Since the surds form a field, the coefficients in this equation are all surds. If $x_1 = x_2$, then $x = x_1$ is an equation of the line.

In both cases, we have shown that an equation for the line through the points (x_1, y_1) and (x_2, y_2) can be expressed in the form $ax + by = c$, where a, b, and c are all surds and a and b are not both 0. $\square$

Theorem 12.3.8. *A circle whose center is in the surd plane and whose radius is a surd has an equation in which the coefficients are all surds.*

Proof. Let the center be (x_1, y_1) and the radius be r. Then one equation of the circle is $(x - x_1)^2 + (y - y_1)^2 = r^2$. Expanding this equation and using the fact that the set of surds is a field shows that this equation has surd coefficients. $\square$

Theorem 12.3.9. *The point of intersection of two distinct nonparallel lines that each go through two points in the surd plane is itself in the surd plane.*

Proof. By Theorem 12.3.7, each of the lines has an equation with surd coefficients. Let such equations be $a_1x + b_1y = c_1$ and $a_2x + b_2y = c_2$. If $a_1 = 0$, then $a_2 \neq 0$ (or else the two lines would be parallel). Then $y = \frac{c_1}{b_1}$, so $a_2x + b_2\frac{c_1}{b_1} = c_2$ from which it follows that the intersection of the two lines has coordinates $x = \frac{c_2}{a_2} - \frac{b_2}{a_2}\frac{c_1}{b_1}$ and $y = \frac{c_1}{b_1}$, both of which are surds.

If $a_1 \neq 0$, then $x = -\frac{b_1}{a_1}y + \frac{c_1}{a_1}$. Substituting this in the second equation yields $a_2\left(-\frac{b_1}{a_1}y + \frac{c_1}{a_1}\right) + b_2y = c_2$. Since the coefficients are all surds, it is clear that y is also a surd. Hence, so is x and the theorem is proven in this case as well.   □

We next consider the points of intersection of a line and a circle.

Theorem 12.3.10. *The points of intersection of a line that has an equation with surd coefficients and a circle that has an equation with surd coefficients lie in the surd plane.*

Proof. Consider a line with equation $ax + by = c$ and a circle with equation $(x - f)^2 + (y - g)^2 = r^2$, where all of the coefficients are surds. Consider first the case where $a = 0$. In this case, $y = \frac{c}{b}$. Substituting this in the equation of the circle yields $(x - f)^2 + (\frac{c}{b} - g)^2 = r^2$. This is a quadratic equation in x. It has 0, 1, or 2 real number solutions depending upon whether the line does not intersect the circle, is tangent to the circle, or intersects the circle in two points. The quadratic formula shows that solutions that exist are obtained from the coefficients by the ordinary arithmetic operations and the extracting of a square root. All of these operations on surds produce surds. Thus, any solutions x are surds, proving the theorem in this case.

If $a \neq 0$, then $x = -\frac{b}{a}y + \frac{c}{a}$. Substituting this value in the equation of the circle yields $\left(-\frac{b}{a}y + \frac{c}{a} - f\right)^2 + (y - g)^2 = r^2$. As above, any solutions of this equation are also surds. Therefore, the theorem holds in this case too.   □

The remaining case is the intersection of two circles.

Theorem 12.3.11. *The points of intersection of two distinct circles that have equations with surd coefficients lie in the surd plane.*

Proof. In order for two distinct circles to intersect, they must have distinct centers. Thus, the equations of the circles can be written in the form

$$(x - a_1)^2 + (y - b_1)^2 = r_1^2$$

$$(x - a_2)^2 + (y - b_2)^2 = r_2^2$$

or

$$x^2 - 2a_1x + a_1^2 + y^2 - 2b_1y + b_1^2 = r_1^2$$

$$x^2 - 2a_2x + a_2^2 + y^2 - 2b_2y + b_2^2 = r_2^2$$

where (a_1, b_1) and (a_2, b_2) are distinct points. (This means that $a_1 \neq a_2$ or $b_1 \neq b_2$.) Subtracting the second equation from the first shows that any point (x, y) that lies on both circles also lies on the line with equation

$$(-2a_1 + 2a_2)x + a_1^2 - a_2^2 + (-2b_1 + 2b_2)y + b_1^2 - b_2^2 = r_1^2 - r_2^2$$

Since this equation has surd coefficients, all points of intersection of this line with either circle lie in the surd plane (Theorem 12.3.10). □

Theorem 12.3.12. *The field of constructible numbers is the same as the field of surds.*

Proof. We already showed that every surd is constructible (Theorem 12.3.3). On the other hand, Theorems 12.3.9, 12.3.10, and 12.3.11 show that the only constructible points in the plane are points with surd coordinates. Since every constructible number is a coordinate of a constructible point in the plane (Theorem 12.3.5), it follows that every constructible number is a surd. □

This characterization of the constructible numbers is the key to the proof that certain angles cannot be trisected. One of the relationships between constructible angles and constructible numbers is the following; we restrict the discussion to acute angles (i.e., angles less than a right angle) simply to avoid having to describe several cases.

Theorem 12.3.13. *The acute angle θ is constructible with a straightedge and compass if and only if $\cos \theta$ is a constructible number.*

Proof. Suppose first that the angle θ is constructible. Place the angle so that its vertex lies at the point 0 on the number line, one of its sides is the positive part of the number line and the other side is on top of it, as in Figure 12.10. Use the compass to mark a point on the upper side of the angle that is one unit from the point 0. Drop a perpendicular from that point to the number line (as in the proof of Theorem 12.3.5). Then that perpendicular meets the number line at $\cos \theta$, so $\cos \theta$ is constructed.

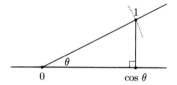

Fig. 12.10 Constructing the cosine of an angle

Conversely, if $\cos \theta$ is constructed, erect a perpendicular upwards from the point $\cos \theta$ on the number line. Construct the number $a = \sqrt{1 - \cos^2 \theta}$ and mark the point on the perpendicular with that distance above the number line, as in

Figure 12.11. Connecting the point 0 to that marked point by a straightedge produces the angle θ.   □

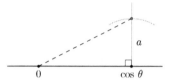

a

0 $\cos\theta$

Fig. 12.11 Constructing an angle from its cosine

With this background we can now determine exactly which angles with an integral number of degrees are constructible. First note the following.

Theorem 12.3.14. *An angle of 60° is constructible.*

Proof. This is an immediate consequence of Theorem 12.3.13, for the cosine of 60° equals $\frac{1}{2}$, and $\frac{1}{2}$ is a constructible number.

There is also an easy direct proof: simply construct an equilateral triangle using a straightedge and compass; each angle of the equilateral triangle is 60°.

To construct an equilateral triangle, draw a circle of any radius, call it r, centered at a point A. Draw a line through A that intersects the circle and label a point of intersection B, as in Figure 12.12 below. Next draw a circle of radius r centered at B and label a point of intersection of the two circles with C. Now draw the segments AC and BC. Triangle ABC is an equilateral triangle (all of whose sides have length r).   □

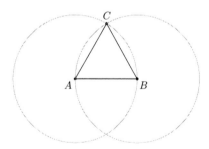

Fig. 12.12 Constructing an equilateral triangle

Corollary 12.3.15. *The following angles are all constructible: 30°, 15°, 45°, and 75°.*

Proof. We begin with the fact that an angle of 60° is constructible (Theorem 12.3.14). An angle of 30° can be constructed by bisecting an angle of 60°, and an angle of 15° can be constructed by bisecting an angle of 30° (Theorem 12.1.4).

An angle of 45° can be constructed by placing an angle of 15° next to one of 30°, and an angle of 75° can be constructed by placing an angle of 15° next to an angle of 60° (Theorem 12.1.7).   □

The material about constructible numbers was developed primarily to prove that some angles are not constructible. We need some additional preliminary results.

Theorem 12.3.16. *For any angle* θ, $\cos(3\theta) = 4\cos^3\theta - 3\cos\theta$.

Proof. Recall the addition formulae for cosine and sine:

$$\cos(\theta_1 + \theta_2) = \cos\theta_1\cos\theta_2 - \sin\theta_1\sin\theta_2$$

and

$$\sin(\theta_1 + \theta_2) = \sin\theta_1\cos\theta_2 + \sin\theta_2\cos\theta_1$$

In particular, if $\theta = \theta_1 = \theta_2$, then

$$\cos(2\theta) = \cos^2\theta - \sin^2\theta$$

and

$$\sin(2\theta) = 2\sin\theta\cos\theta$$

Therefore,

$$\begin{aligned}
\cos(3\theta) &= \cos(2\theta + \theta) \\
&= \cos(2\theta)\cos\theta - \sin(2\theta)\sin\theta \\
&= (\cos^2\theta - \sin^2\theta)\cos\theta - 2\sin\theta\cos\theta\sin\theta \\
&= \cos^3\theta - \sin^2\theta\cos\theta - 2\sin^2\theta\cos\theta \\
&= \cos^3\theta - 3\sin^2\theta\cos\theta
\end{aligned}$$

The trigonometric identity $\sin^2\theta + \cos^2\theta = 1$ implies that $\sin^2\theta = 1 - \cos^2\theta$, which gives

$$\begin{aligned}
\cos(3\theta) &= \cos^3\theta - 3(1 - \cos^2\theta)\cos\theta \\
&= \cos^3\theta - 3\cos\theta + 3\cos^3\theta \\
&= 4\cos^3\theta - 3\cos\theta
\end{aligned}$$

Therefore, $\cos(3\theta) = 4\cos^3\theta - 3\cos\theta$.   □

The case where θ equals 20° is of particular interest.

Corollary 12.3.17. *If* $x = 2\cos(20°)$, *then* $x^3 - 3x - 1 = 0$.

Proof. Using the formula for $\cos(3\theta)$ given above and the fact that the cosine of 60° equals $\frac{1}{2}$, we have $\frac{1}{2} = 4\cos^3(20°) - 3\cos(20°)$. This is equivalent to the equation

$$8\cos^3(20°) - 6\cos(20°) - 1 = 0$$

Since $x = 2\cos(20°)$, $x^3 - 3x - 1 = 0$. $\square$

We will show that the cubic equation $x^3 - 3x - 1 = 0$ does not have a constructible root.

Theorem 12.3.18. *If the roots of the cubic equation* $x^3 + bx^2 + cx + d = 0$ *are* $r_1, r_2,$ *and* r_3, *then* $b = -(r_1 + r_2 + r_3)$. *(It is possible that two or even three of the roots are the same as each other.)*

Proof. By the Factor Theorem (9.3.6), and the fact that the coefficient of x^3 is 1, the cubic equation is the same as $(x - r_1)(x - r_2)(x - r_3) = 0$. Multiplying out these three factors shows that the coefficient of x^2 is $-(r_1 + r_2 + r_3)$; hence, $b = -(r_1 + r_2 + r_3)$. $\square$

We need the concept of a conjugate for elements of $\mathcal{F}(\sqrt{r})$, analogous to the conjugate of a complex number.

Definition 12.3.19. *If* $a + b\sqrt{r}$ *is an element of* $\mathcal{F}(\sqrt{r})$, *then the* conjugate *of* $a + b\sqrt{r}$, *denoted by placing a bar on top of the number, is*

$$\overline{a + b\sqrt{r}} = a - b\sqrt{r}$$

Theorem 12.3.20. *The conjugate of the sum of two elements of* $\mathcal{F}(\sqrt{r})$ *is the sum of the conjugates, and the conjugate of the product of two elements of* $\mathcal{F}(\sqrt{r})$ *is the product of the conjugates.*

Proof. For the first assertion simply note that

$$\begin{aligned}
\overline{(a + b\sqrt{r}) + (c + d\sqrt{r})} &= \overline{(a + c) + (b + d)\sqrt{r}} \\
&= (a + c) - (b + d)(\sqrt{r}) \\
&= (a - b\sqrt{r}) + (c - d\sqrt{r}) \\
&= \overline{(a + b\sqrt{r})} + \overline{(c + d\sqrt{r})}
\end{aligned}$$

For products, note that

$$\begin{aligned}
\overline{(a + b\sqrt{r})(c + d\sqrt{r})} &= \overline{(ac + rbd) + (ad + bc)\sqrt{r}} \\
&= (ac + rbd) - (ad + bc)\sqrt{r}
\end{aligned}$$

and

$$\overline{(a + b\sqrt{r}) \cdot (c + d\sqrt{r})} = (a - b\sqrt{r})(c - d\sqrt{r})$$
$$= (ac + bdr) - (ad + bc)\sqrt{r}$$

Therefore, $\overline{(a + b\sqrt{r})(c + d\sqrt{r})} = \overline{(a + b\sqrt{r})} \cdot \overline{(c + d\sqrt{r})}$.   □

Theorem 12.3.21. *If $a + b\sqrt{r}$ is in $\mathcal{F}(\sqrt{r})$ and is a root of a polynomial with rational coefficients, then $a - b\sqrt{r}$ is also a root of the polynomial.*

Proof. Suppose that $a_n(a+b\sqrt{r})^n + a_{n-1}(a+b\sqrt{r})^{n-1} + \cdots + a_1(a+b\sqrt{r}) + a_0 = 0$. Then,

$$\overline{a_n(a + b\sqrt{r})^n + a_{n-1}(a + b\sqrt{r})^{n-1} + \cdots + a_1(a + b\sqrt{r}) + a_0} = 0$$

Since each of the coefficients a_k is rational, $\overline{a_k} = a_k$, for every k. Using this fact and Theorem 12.3.20 (the facts that the conjugate of a sum is the sum of the conjugates and the conjugate of a product is the product of the conjugates), it follows that $a_n\overline{(a + b\sqrt{r})}^n + a_{n-1}\overline{(a + b\sqrt{r})}^{n-1} + \cdots + a_1\overline{(a + b\sqrt{r})} + a_0 = 0$. Thus, $a - b\sqrt{r} = \overline{a + b\sqrt{r}}$ is also a root of the polynomial.   □

Theorem 12.3.22. *If a cubic equation with rational coefficients has a constructible root, then the equation has a rational root.*

Proof. Dividing through by the leading coefficient if necessary, we can assume that the coefficient of x^3 is 1. Then, by Theorem 12.3.18, the sum of the three roots of the cubic equation is rational.

We first show that if the equation has a root in any $\mathcal{F}(\sqrt{r})$, then it has a root in $\mathcal{F}$. To see this, suppose the equation has a root in $\mathcal{F}(\sqrt{r})$ of the form $a + b\sqrt{r}$ with $b \neq 0$. Then, by Theorem 12.3.21, the conjugate $a - b\sqrt{r}$ is also a root. If r_3 is the third root and s is the sum of all three roots, then $s = r_3 + (a+b\sqrt{r}) + (a-b\sqrt{r}) = r_3 + 2a$. Thus, $r_3 = s - 2a$. Since $\mathcal{F}$ contains all rational numbers and s is rational, s is in $\mathcal{F}$. Since a is also in $\mathcal{F}$, it follows that the root r_3 is in $\mathcal{F}$ itself.

The preliminary result obtained in the previous paragraph allows us to prove the theorem as follows. If the polynomial has a constructible root, then, since every constructible number is a surd (Theorem 12.3.12), the root is in a field that occurs at the end of a tower. Consider the field at the end of the shortest tower that contains any root of the given cubic equation. We claim that field must be $\mathbb{Q}$. To see this, simply note that if the root was in an $\mathcal{F}(\sqrt{r})$, the previous paragraph would imply that the root was in $\mathcal{F}$, which would be at the end of a shorter tower than $\mathcal{F}(\sqrt{r})$ is. Hence, that root must be in $\mathbb{Q}$. Thus, the equation has a rational root.   □

We can now prove that an angle of 20° cannot be constructed.

Theorem 12.3.23. *An angle of 20° cannot be constructed with straightedge and compass.*

Proof. If an angle of 20° could be constructed with straightedge and compass, then $\cos(20°)$ would be a constructible number (Theorem 12.3.13). Then $2\cos(20°)$ would also be a constructible number, and the polynomial $x^3 - 3x - 1 = 0$ would therefore have a constructible root (Corollary 12.3.17). It follows from the previous theorem (12.3.22) that this polynomial would need to have a rational root. Thus, to establish that an angle of 20° is not constructible, all that remains to be shown is that the polynomial $x^3 - 3x - 1 = 0$ does not have a rational root. This can be proven as an application of the Rational Roots Theorem (8.1.9). However, to make the present result independent of that theorem, we present a direct proof.

Suppose that m and n are integers with $n \neq 0$ and that $\frac{m}{n}$, written in lowest terms, is a root of the equation $x^3 - 3x - 1 = 0$. Then $\frac{m^3}{n^3} - 3\left(\frac{m}{n}\right) - 1 = 0$ implies that $m^3 - 3mn^2 - n^3 = 0$. Since $n^3 = m(m^2 - 3n^2)$, every prime number dividing m also divides n^3 and hence also divides n (Corollary 4.1.3). Since m and n are relatively prime, there are no primes that divide m. Thus, m is either 1 or -1. Similarly, since $m^3 = n(3mn + n^2)$, any prime that divides n also divides m, from which it follows that n is 1 or -1. Hence, $\frac{m}{n}$ is 1 or -1. Therefore, the only possible rational roots of $x^3 - 3x - 1 = 0$ are $x = 1$ or $x = -1$. Substituting those values for x in the equation shows that neither of those is a root, so the theorem is proven.   □

Corollary 12.3.24. *An angle of 60° cannot be trisected with straightedge and compass.*

Proof. As we have seen, an angle of 60° can be constructed with a straightedge and compass (Theorem 12.3.14). If an angle of 60° could be trisected with straightedge and compass, then an angle of 20° would be constructible. But an angle of 20° is not constructible, by the previous theorem (12.3.23).   □

12.4   Constructions of Geometric Figures

Another problem that the Ancients Greeks raised but could not solve was what they called *duplication of the cube*. This was the question of whether or not a side of a cube of volume 2 could be constructed by straightedge and compass.

Theorem 12.4.1. *The side of a cube of volume 2 cannot be constructed with a straightedge and compass.*

Proof. If x is the length of the side of a cube of volume 2, then, of course, $x^3 = 2$, or $x^3 - 2 = 0$. By Theorem 12.3.22, this equation has a constructible root if and only if it has a rational root. Since the cube root of 2 is irrational (Problem 13 in Chapter 8), there is no constructible solution, and the cube cannot be "duplicated" using only a straightedge and compass.   □

The question of which regular polygons can be constructed is very interesting.

Definition 12.4.2. A *polygon* is a figure in the plane consisting of line segments that bound a finite portion of the plane. A *regular polygon* is a polygon all of whose angles are equal and all of whose sides are equal.

An equilateral triangle is a regular polygon with three sides. Equilateral triangles can easily be constructed with straightedge and compass (see the proof of Theorem 12.3.14).

A square is a regular polygon with four sides. It is also very easy to construct a square. Simply use the straightedge to draw any line segment, and erect perpendiculars at each end of the line segment. Then use the compass to "measure" the length of the line segment and mark points which are that distance above the original line segment on each of the perpendiculars. Using the straightedge to connect those points yields a square.

For each natural number n bigger than or equal to 3, there exists a regular polygon with n equal sides. This can be seen as follows. (Which of these regular polygons is constructible is a more difficult question that we discuss in Theorem 12.4.5.)

Theorem 12.4.3. *For each natural number n greater than or equal to 3 there is a regular polygon with n sides inscribed in a circle.*

Proof. Given a natural number n bigger than or equal to 3, take a circle and draw successive adjacent angles of $\frac{360}{n}$ degrees at the center, as shown in Figure 12.13. Then draw the line segments connecting adjacent points determined by the sides of the angles intersecting the circumference of the circle. We must show that those line segments are all equal in length and that the angles formed by each pair of adjacent line segments are equal to each other.

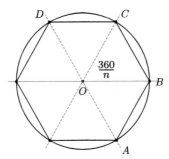

Fig. 12.13 Existence of regular polygons

Consider, for example, the triangles OAB and OCD in Figure 12.13. The angles AOB and COD are each equal to $\frac{360}{n}$ degrees. The sides OA, OB, OC, and OD are all radii of the given circle and are therefore equal to each other. It follows that $\triangle OAB$ is congruent to $\triangle OCD$ by side-angle-side (11.1.2). The same proof shows that all of the triangles constructed are congruent to each other. It follows that all of the sides of the polygon, which are the sides opposite the angles of $\frac{360}{n}$ degrees in the triangles, are equal to each other. The angles of the polygon are angles such as $\angle ABC$ and $\angle BCD$ in the diagram. Each of them is the sum of two base angles of the drawn triangles, and, therefore, the angles of the polygon are equal to each other as well.   □

Definition 12.4.4. A *central angle* of a regular polygon with n sides is the angle of $\frac{360}{n}°$ that has a vertex at the center of the polygon, as in the above proof.

Theorem 12.4.5. *A regular polygon is constructible if and only if its central angle is a constructible angle.*

Proof. Suppose that a regular polygon can be constructed with straightedge and compass. Then its center (a point equidistant from all of its vertices) can be constructed as the point of intersection of the perpendicular bisectors of two adjacent sides of the polygon (see Problem 13 at the end of this chapter). Now the central angle can be constructed as the angle formed by connecting the center to two adjacent vertices of the polygon. All such angles are equal to each other, since the corresponding triangles are congruent by side-side-side (11.1.8). There are n such angles, the sum of which is 360 degrees, so each central angle is $\frac{360}{n}°$.

Conversely, suppose that an angle of $\frac{360}{n}°$ is constructible, for some natural number $n \geq 3$. Then a regular polygon with n sides can be constructed as follows. Make a circle. Construct an angle of $\frac{360}{n}°$ with vertex at the center of the circle. Then construct another such angle adjacent to the first, and so on until n such angles have been constructed. Connecting the adjacent points of intersection of the sides of those angles with the circle constructs a regular polygon with n sides (as shown in the proof of Theorem 12.4.3).   □

Corollary 12.4.6. *A regular polygon with 18 sides cannot be constructed with a straightedge and compass.*

Proof. A regular polygon with 18 sides has a central angle of $\frac{360}{18} = 20$ degrees. We proved in Theorem 12.3.23 that an angle of $20°$ is not constructible, so the previous theorem implies that a regular polygon with 18 sides is not constructible.   □

Theorem 12.4.7. *If m is a natural number greater than 2, then a regular polygon with $2m$ sides is constructible if and only if a regular polygon with m sides is constructible.*

Proof. Using Theorem 12.4.5, the result follows by either bisecting or doubling the central angle of the already constructed polygon. (Alternatively, having constructed a regular polygon with $2m$ sides, use the straightedge to connect alternate vertices, yielding a regular polygon with m sides, as can be established by using congruent triangles. In the other direction, given a regular polygon with m sides, inscribe it in a circle and then double the vertices by adding the points of intersections of the perpendicular bisectors of the sides and the circle.)   □

Corollary 12.4.8. *A regular polygon with 9 sides is not constructible.*

Proof. This follows immediately from the fact that a regular polygon with 18 sides is not constructible (Corollary 12.4.6) and the above theorem (12.4.7).   □

It is useful to make the following connection between constructible polygons and constructible numbers.

Theorem 12.4.9. *A regular polygon with n sides is constructible if and only if the length of the side of a regular polygon with n sides that is inscribed in a circle of radius 1 is a constructible number.*

Proof. In the first direction suppose that a regular polygon with n sides is constructible. Then such a polygon can be constructed so that it is inscribed in a circle of radius 1 (for example, by putting its constructible central angle in a circle of radius 1). The length of the side can be constructed by using the compass to "measure" the side of the constructed polygon.

Conversely, if s is a constructible number and is the length of the side of a regular polygon with n sides inscribed in a circle of radius 1, then the regular polygon can be constructed simply by marking any point on the circle and then using the compass to successively mark points that are at distance s from the last marked one. The marked points will be vertices of a regular polygon with n sides. □

Can a pentagon (a regular polygon with 5 sides) be constructed using only a straightedge and compass? The answer is affirmative, but it is not at all easy to see directly. We will approach this by considering a regular polygon with 10 sides.

Theorem 12.4.10. *A regular polygon with 10 sides is constructible.*

Proof. By Theorem 12.4.9, it suffices to show that the length of a side of such a polygon inscribed in a circle of radius 1 is a constructible number. We determine the length of such a side by using a little geometry. The central angle of a regular polygon with 10 sides is $36°$. Consider such an angle with vertex O at the center of a circle of radius 1, as shown in Figure 12.14. Label the points of intersection of the sides of that central angle with the circle A and B. Let s denote the length of the line segment from A to B, and let AC be the bisector of $\angle OAB$. Since $\angle OAB$ is $72°$ (the sum of the degrees of the equal angles OAB and ABO must be $180° - 36°$), it follows that angles OAC and CAB are each $36°$. Also, $\angle OBA$ is $72°$. Thus, triangles OAB and CAB are similar to each other, so corresponding sides are in proportion (Theorem 11.3.11). Therefore, triangle CAB is isosceles, and AC has length s.

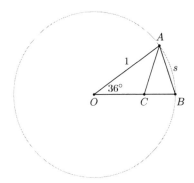

Fig. 12.14 The side of a ten-sided regular polygon

Since $\angle AOB = 36° = \angle OAC$, $\triangle OAC$ is also isosceles. Thus, OC has length s, from which it follows that BC has length $1 - s$. The side opposite the $36°$ angle in $\triangle OAB$, with length s, is to the side opposite the $36°$ angle of $\triangle CAB$, with length $1 - s$, as the side opposite the $72°$ angles of $\triangle OAB$, with length 1 (the radius of the circle), is to the side opposite the $72°$ angle of $\triangle CAB$, which has length s. That is,

$$\frac{s}{1-s} = \frac{1}{s}$$

Thus, the length we are interested in, s, satisfies the equation $s^2 = 1 - s$, or $s^2 + s - 1 = 0$. The positive solution of this equation (s is a length) is $\frac{-1+\sqrt{5}}{2}$. Thus, s is a constructible number (Theorem 12.3.12), from which it follows that the regular polygon with 10 sides is constructible.   □

Corollary 12.4.11. *A regular pentagon is constructible.*

Proof. This follows immediately from the above theorem and Theorem 12.4.7.   □

Which regular polygons are constructible? Those with 3, 4, and 5 sides are, and thus, so are 6, 8, and 10 (Theorem 12.4.7). We proved that a regular polygon with 9 sides is not constructible (Corollary 12.4.8).

What about a polygon with 7 sides? We can approach this question using some facts that we learned about complex numbers. As follows immediately from a previous result (Example 9.2.11), for each natural number n greater than 2, the complex solutions to the equation $z^n = 1$ are the vertices of an n-sided regular polygon inscribed in a circle of radius 1. We will approach the problem by considering the solutions of $z^7 = 1$.

Theorem 12.4.12. *A regular polygon with 7 sides is not constructible.*

Proof. If a regular polygon with 7 sides was constructible, then one could be constructed inscribed in a circle of radius 1 centered at the origin, such that one of the vertices lies on the x-axis at the point corresponding to the number 1. Then the vertices are the 7th-roots of unity (Example 9.2.11); that is, they satisfy $z^7 = 1$.

We will analyze the first vertex above the x-axis. Let that vertex lie at the complex number z_0. If the regular polygon was constructible, then z_0 would be a constructible point, and therefore the real part of z_0 would be constructible (simply construct a perpendicular from z_0 to the x-axis, as described in the parenthetical remark at the end of the proof of Theorem 12.3.5). It would follow that twice the real part is constructible. Let x_0 be twice that real part. We will show that x_0 satisfies a cubic equation that is not satisfied by any constructible number.

Begin by observing that $x_0 = z_0 + \overline{z_0}$. Since $|z_0| = 1$, it follows that $1 = |z_0|^2 = z_0\overline{z_0}$. Thus, $\overline{z_0} = \frac{1}{z_0}$, so $x_0 = z_0 + \frac{1}{z_0}$. The cubic equation satisfied by x_0 will be obtained from the equation of degree 7 satisfied by z_0, $z_0^7 = 1$, and the fact that $z_0 \neq 1$. Note that $z^7 - 1 = (z-1)(z^6 + z^5 + z^4 + z^3 + z^2 + z + 1)$. Since $z_0 - 1 \neq 0$, $z_0^6 + z_0^5 + z_0^4 + z_0^3 + z_0^2 + z_0 + 1 = 0$. Dividing through by z_0^3 yields

$$z_0^3 + z_0^2 + z_0 + 1 + \frac{1}{z_0} + \frac{1}{z_0^2} + \frac{1}{z_0^3} = 0$$

Note that $\left(z_0 + \frac{1}{z_0}\right)^3 = z_0^3 + 3z_0 + \frac{3}{z_0} + \left(\frac{1}{z_0}\right)^3$ and also that $\left(z_0 + \frac{1}{z_0}\right)^2 = z_0^2 + 2 + \left(\frac{1}{z_0}\right)^2$. It follows that

$$z_0^3 + z_0^2 + z_0 + 1 + \frac{1}{z_0} + \frac{1}{z_0^2} + \frac{1}{z_0^3} = \left(z_0 + \frac{1}{z_0}\right)^3 + \left(z_0 + \frac{1}{z_0}\right)^2 - 2\left(z_0 + \frac{1}{z_0}\right) - 1$$

Then, since $x_0 = z_0 + \frac{1}{z_0}$, x_0 satisfies the equation

$$x_0^3 + x_0^2 - 2x_0 - 1 = 0$$

As indicated, to show that a regular polygon with 7 sides is not constructible, it suffices to show that x_0 is not a constructible number. Since x_0 satisfies this cubic equation with rational coefficients, the result will follow if it is shown that this cubic equation has no rational root (Theorem 12.3.22). We could use the Rational Roots Theorem (8.1.9). Alternatively, suppose that the rational number $\frac{m}{n}$ satisfied this cubic equation. We can, and do, assume that m and n have no common integral factor other than 1 and -1. Then $\left(\frac{m}{n}\right)^3 + \left(\frac{m}{n}\right)^2 - 2\left(\frac{m}{n}\right) - 1 = 0$, or $m^3 + m^2 n - 2mn^2 - n^3 = 0$. Now if p was a prime number that divided m, it would follow from the above that p would divide n^3 and hence also divide n. Since m and n are relatively prime there is no such prime number p, and we conclude that m is either 1 or -1. Similarly, n is equal to 1 or n is equal to -1. Thus, $\frac{m}{n}$ equals 1 or -1. But $1^3 + 1^2 - 2 - 1$ is not 0, nor is $(-1)^3 + (-1)^2 + 2 - 1$. Hence, there is no rational solution, and the theorem is proven.   □

It is known exactly which regular polygons are constructible. The Gauss-Wantzel Theorem states that a regular polygon with n sides is constructible if and only if n is 2^k, where $k > 1$, or $2^k F_1 \cdots F_l$, where $k \geq 0$ and the F_j are distinct Fermat primes. Recall (Problem 14 in Chapter 2) that a Fermat number is a number of the form $2^{2^n} + 1$ for nonnegative integers n. A *Fermat prime* is a Fermat number that is prime. The first few Fermat numbers are 3 (when $n = 0$), 5 (when $n = 1$), 17 (when $n = 2$), and 257 (when $n = 3$). Fermat thought that all Fermat numbers might be prime, but Euler found that the fifth Fermat number is not prime. It is a remarkable fact that it is unknown whether or not there are an infinite number of Fermat primes. (It is equally remarkable that it is not known whether there are infinitely many composite Fermat numbers.) It is therefore not known whether or not there are an infinite number of constructible regular polygons with an odd number of sides.

We can determine exactly which angles having a natural number of degrees are constructible.

Theorem 12.4.13. *If n is a natural number, then an angle of n degrees is constructible if and only if n is a multiple of 3.*

Proof. Recall that we proved that a regular polygon with 10 sides is constructible (Theorem 12.4.10) and, hence, that an angle of 36° is constructible (Theorem 12.4.5). Since an angle of 30° is constructible (Corollary 12.3.15), we can "subtract" a 30° angle from a 36° angle by placing the 30° angle with the vertex and one of its sides coincident with the vertex and one of the sides of the 36° angle (Theorem 12.1.6). Then, bisecting the constructed angle of 6° yields an angle of 3°. Once an angle of 3° is constructed, an angle of $3k$ degrees can be constructed by simply placing k angles of 3° appropriately adjacent to each other.

To establish the converse, suppose that an angle of n degrees is constructible. We must show that n is congruent to 0 (mod 3). If n was congruent to either 1 or 2 modulo 3, then we could construct an angle of 1° or 2° accordingly by "subtracting" an appropriate number of angles of 3° from the angle of n degrees. If the resulting angle is 2°, bisecting it would yield an angle of 1°. Thus, if an angle of n degrees was constructible and n was not a multiple of 3, then an angle of 1° could be constructed. But an angle of 1° is not constructible, for if it was, placing 20 of them together would contradict the fact that an angle of 20° is not constructible (12.3.23).   □

We have shown that some angles, such as an angle of 60°, cannot be trisected with a straightedge and compass. But what about the following?

Example 12.4.14 (Trisection of arbitrary acute angles). Let θ be any acute angle. Mark any two points on your straightedge and let the distance between them be r. Draw the angle θ and construct the circle with radius r whose center is at the vertex of θ. Label the center of the circle O. Extend one of the sides of θ in both directions. Move the marked straightedge so that the point marked to the left is on the extended line, the point marked to the right stays on the circle, and the straightedge passes through the intersection of the circle and the side of $\angle \theta$ that was not extended; label the points of intersection A, B, C, as shown in Figure 12.15. Draw the line BO. Then the line segments AB, BO, and OC all have length r. Now let the equal base

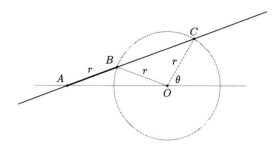

Fig. 12.15 On the way to trisecting an arbitrary angle

angles of $\triangle ABO$ be x, the equal base angles of $\triangle OBC$ be y, and let $\angle BOC$ be z, as shown in Figure 12.16. Then the sum of $\angle ABO$ and $2x$ is 180°, and the sum of $\angle ABO$ and y is also $180°$; hence $y = 2x$. It is clear that $x + z + \theta$ is $180°$. On the other hand, $z + 2y$ is $180°$. Since $y = 2x$, $4x + z$ is also $180°$. It follows that $4x + z = x + \theta + z$, or $3x = \theta$. Thus, the angle x is one third of θ, so θ has been trisected.   □

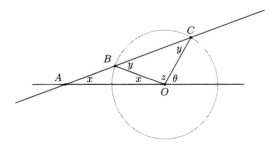

Fig. 12.16 Trisecting an arbitrary angle

What is going on here? You may think that the construction we have just done contradicts our earlier proof that an angle of $60°$ cannot be trisected. However, the construction in the example above violated the classical rules of constructions that we were adhering to before this example. Namely, we marked two points on the straightedge. What we have shown is that it is possible to trisect arbitrary angles with a compass and straightedge on which two (or more) points are marked. Therefore, in particular, any angle can be trisected using a *ruler* and compass, but not with a mere straightedge and compass.

12.5   Problems

Basic Exercises

1. Determine which of the following numbers are constructible:

(a) $\dfrac{1}{\sqrt{3+\sqrt{2}}}$

(b) $\sqrt[6]{79}$

(c) 3.146891

(d) $\sqrt[16]{79}$

(e) $\sqrt{6 + \frac{\sqrt[3]{4}}{2}}$

(f) $\sqrt{7 + \sqrt{5}}$

(g) $\sqrt{3 + 4\sqrt{2} + \sqrt{5}}$

(h) $\sqrt[3]{\frac{9}{10}}$

(i) $\sqrt{\dfrac{3792}{1419}}$

(j) $\cos 51°$

(k) $\cos 5°$

(l) $\cos 10°$

(m) $11^{\frac{2}{3}}$

(n) $11^{\frac{3}{2}}$

(o) $2^{\frac{1}{6}}$

(p) $2^{\frac{3}{2}}$

(q) $\sqrt[3]{\frac{\sqrt{2}}{4}}$

(r) $\sqrt{7}\cos 15°$

2. Determine which of the following angles are constructible:

(a) 6°   (f) 15°   (j) 37.5°
(b) 5°   (g) 75°   (k) 7.5°
(c) 10°   (h) 80°   (l) 120°
(d) 30°   (i) 92.5°   (m) 160°
(e) 35°

3. Determine which of the following angles can be trisected:

(a) 12°
(b) 30°

Interesting Applications

4. Determine which of the following polynomials have at least one constructible root:

(a) $x^4 - 3$ (f) $x^3 - 2x - 1$
(b) $x^8 - 7$ (g) $x^3 + 4x + 1$
(c) $x^4 + \sqrt{7}x^2 - \sqrt{3} - 1$ (h) $x^3 + 2x^2 - x - 1$
(d) $x^3 + 6x^2 + 9x - 10$ (i) $x^3 - x^2 + x - 1$
(e) $x^3 - 3x^2 - 2x + 6$ (j) $2x^3 - 4x^2 + 1$

5. Determine which of the following regular polygons can be constructed with straightedge and compass:

(a) A regular polygon with 14 sides
(b) A regular polygon with 20 sides
(c) A regular polygon with 36 sides
(d) A regular polygon with 240 sides

6. Explain how to construct a regular polygon with 24 sides using straightedge and compass.

7. True or False:

(a) If the angle of θ degrees is constructible and the number x is constructible, then the angle of $x \cdot \theta$ degrees is constructible.
(b) x^y is constructible if x and y are each constructible.
(c) If $\frac{x}{z}$ is constructible, then x and z are each constructible.
(d) There is an angle θ such that $\cos \theta$ is constructible, but $\sin \theta$ is not constructible.

8. For an acute angle θ, show that $\tan \theta$ is a constructible number if and only if θ is a constructible angle.

9. Determine which of the following numbers are constructible:

   (a) $\sin 20°$
   (b) $\sin 75°$
   (c) $\tan 2.5°$

10. Determine which of the following numbers are constructible (the angles below are in radians):

    (a) $\sin \frac{\pi}{16}$
    (b) $\cos \pi$
    (c) $\tan \frac{\pi}{4}$

11. (a) Prove that the cube cannot be tripled, in the sense that, starting with an edge of a cube of volume 1, an edge of a cube of volume 3 cannot be constructed with straightedge and compass.
    (b) More generally, prove that the side of a cube with volume a natural number n is constructible if and only if $n^{\frac{1}{3}}$ is a natural number.

12. Using mathematical induction, prove that, for every integer $n \geq 1$, a regular polygon with $3 \cdot 2^n$ sides can be constructed with straightedge and compass.

Challenging Problems

13. Prove that, given a regular polygon, its center can be constructed using only a straightedge and compass.
    [Hint: The center can be determined as the point of intersection of the perpendicular bisectors of two adjacent sides of the polygon. To prove that this point is indeed the center, prove that all the right triangles with one side a perpendicular bisector of a side of the polygon, another side a half of a side of the polygon, and the third side the line segment joining the "center" to a vertex of the polygon are congruent to each other.]

14. Prove that an acute angle cannot be trisected with straightedge and compass if its cosine is:

    (a) $\frac{3}{7}$ (d) $\frac{3}{5}$
    (b) $\frac{2}{5}$ (e) $\frac{1}{4}$
    (c) $\frac{1}{5}$

15. Can a polynomial of degree 4 with rational coefficients have a constructible root without having a rational root?

16. Prove that the following equation has no constructible solutions:

$$x^3 - 6x + 2\sqrt{2} = 0$$

    [Hint: You can use Theorem 12.3.22 if you make an appropriate substitution.]

17. Let t be a transcendental number. Prove that $\{(a + bt) : a, b \in \mathbb{Q}\}$ is not a subfield of $\mathbb{R}$.

18. Say that a complex number $a + bi$ is constructible if the point (a, b) is constructible (equivalently, if a and b are both constructible real numbers). Show that the cube roots of $\frac{1}{2} + \frac{\sqrt{3}}{2}i$ are not constructible.

19. Let $\mathcal{F}$ be the smallest subfield of $\mathbb{R}$ that contains π.

    (a) Show that $\mathcal{F}$ consists of all numbers that can be written in the form $\frac{p(\pi)}{q(\pi)}$, where p and q are polynomials with rational coefficients and q is not the zero polynomial.

    (b) Show that $\mathcal{F}$ is countable.

20. Is $\{a\sqrt{2} : a \in \mathbb{Q}\}$ a subfield of $\mathbb{R}$?

21. Is the set of all towers countable? (Recall that a *tower* is a finite sequence of subfields of $\mathbb{R}$, the first of which is $\mathbb{Q}$, such that the other subfields are obtained from their predecessors by adjoining square roots.)

22. Prove the following:

    (a) If x_0 is a root of a polynomial with coefficients in $\mathcal{F}(\sqrt{r})$, then x_0 is a root of a polynomial with coefficients in $\mathcal{F}$.

    (b) Every constructible number is algebraic.

    (c) The set of constructible numbers is countable.

    (d) There is a circle with center at the origin that is not constructible.

23. Let t be a transcendental number. Prove that t cannot be a root of any equation of the form $x^2 + ax + b = 0$, where a and b are constructible numbers.

24. Is there a line in the plane such that every point on it is constructible?

25. Find the cardinality of each of the following sets:

    (a) The set of roots of polynomials with constructible coefficients

    (b) The set of constructible angles

    (c) The set of all points (x, y) in the plane such that x is constructible and y is irrational

    (d) The set of all sets of constructible numbers

26. (Very challenging) Use a straightedge and compass to directly (without first constructing its central angle or the length of the side of any polygon) construct a regular pentagon.

27. Suppose that regular polygons with m sides and n sides can be constructed and m and n are relatively prime. Prove that a regular polygon of mn sides can be constructed.
    [Hint: Use central angles and use the fact that a linear combination of m and n is 1.]

28. Prove the following: For natural numbers m and n, if a given angle can be divided into n equal parts using only a straightedge and compass, and if m is a divisor of n, then the angle can be divided into m equal parts using only a straightedge and compass.

29. (Very challenging) Prove that you cannot trisect an angle by trisecting the side opposite the angle in a triangle containing it. That is, prove that, if ABC is any triangle, there do not exist two lines through A such that those lines trisect both the side BC of the triangle and the angle BAC of the triangle.

    [Hint: Suppose that there do exist two such lines. The lines then divide the triangle into three sub-triangles. One approach uses the easily established fact that all three sub-triangles have the same area.]

Index

D. Rosenthal et al., *A Readable Introduction to Real Mathematics*,
Undergraduate Texts in Mathematics, DOI 10.1007/978-3-319-05654-8,
© Springer International Publishing Switzerland 2014

CPSIA information can be obtained
at www.ICGtesting.com
Printed in the USA
LVOW10s1025210317

527948LV00010B/264/P